OXFORD MEDICAL PUBLICATIONS

Hypothermia

THE FACTS

Hypothermia

THE FACTS

K. J. COLLINS

Member of the Scientific Staff of the Medical Research Council and
Honorary Senior Lecturer in Geriatric Medicine,
University College Hospital, London

OXFORD   NEW YORK   TORONTO
OXFORD UNIVERSITY PRESS
1983

Oxford University Press, Walton Street, Oxford OX2 6DP
London Glasgow New York Toronto
Delhi Bombay Calcutta Madras Karachi
Kuala Lumpur Singapore Hong Kong Tokyo
Nairobi Dar es Salaam Cape Town
Melbourne Auckland
and associated companies in
Beirut Berlin Ibadan Mexico City Nicosia

OXFORD is a trade mark of Oxford University Press

© K. J. Collins, 1983

British Library Cataloguing in Publication Data

Collins, K.J.
Hypothermia.——(Oxford medical publications)
1. Hypothermia
I. Title
616.9'88   RC103.H/
ISBN 0-19-261360-X

Library of Congress Cataloguing in Publication Data

Collins, K.J. (Kenneth John), 1929-
Hypothermia: the facts.
(Oxford medical publications)
Bibliography: p.
Includes index.
1. Hypothermia—Addresses, essays, lectures.
I. Title.   II. Series.   [DNLM: 1. Hypothermia.   WD 670 C712h]
RC88.5.C64   1983   616.9'89   83-4024
ISBN 0-19-261360-X (U.S.)

Typeset by Colset Private Limited, Singapore
Printed in Great Britain by R. Clay & Co.,
Bungay, Suffolk

Preface

During the last two or three decades, doctors, health workers, and public-spirited laymen throughout the world have become increasingly concerned with hypothermia and the various conditions leading to its development. It has become a topic popularly recognized and discussed in the media, often with emphasis on the dramatic circumstances of its discovery, and it has sometimes been regarded as a political yardstick with which to measure social need. But how common is hypothermia? And, in its urban form, is it truly a peculiarly British complaint, especially of the elderly, reflecting the preponderance of old, poorly insulated, and inadequately heated housing stock in this country?

Hypothermia literally means low body temperature, but though we are now more aware of its occurrence it is not a disorder which especially afflicts contemporary society. The human being is a homeotherm, which means that men and women are able to exist in a wide variety of temperature conditions and yet maintain internal body temperature constant within a narrow range. If, for any reason, body temperature is allowed to fall to low levels, or for that matter to rise to high levels, outside this range, then the body reacts in such a way as to restore normal body temperature again. It is a serious matter if this balance is not achieved; so serious that death may follow if the temperature swing is profound and irreversible. Primitive societies were surely as much prone to this threat as their modern counterparts.

Much present-day concern with hypothermia is with the elderly, and rightly so, for old people can become particularly vulnerable to the effects of cold. The main purpose of this

Preface

book is to provide a perspective of the nature of the problem of hypothermia in old people and to show how it arises, how it can be recognized, and how it can be prevented or treated. Given the right circumstances any one of the seven ages of man may become susceptible to cold-temperature conditions, and the special problems of body-temperature control in the new-born and the hypothermia which can develop in adults as the result of cold exposure or even through the effects of certain drugs need also to be considered.

Facts are established on the basis of available evidence and scientific proof, and it would be a bold man indeed who would claim that any of the known facts about hypothermia will not require reappraisal at some stage. The facts in this book summarize our present knowledge and are presented primarily for the guidance of the general public but also, it is hoped, to provide a background for those in the medical and sociological fields who are concerned with the effects of cold and low body temperature.

In this brief account of hypothermia the author is well aware that there are many contributions from pioneers and research workers in this field which have not been recorded. The vital part each has played is readily acknowledged, especially those of my colleagues Professor A.N.Exton-Smith for his work on the clinical aspects of hypothermia and Professors O.G.Edholm and the late J.S.Weiner on the physiology of thermoregulation. I would also like to take the opportunity of thanking the staff of the Oxford University Press for their expert guidance and help during the preparation of this monograph.

London　　　　　　　　　　　　　　　　　　　　　　　K. J. C.
April 1983

Contents

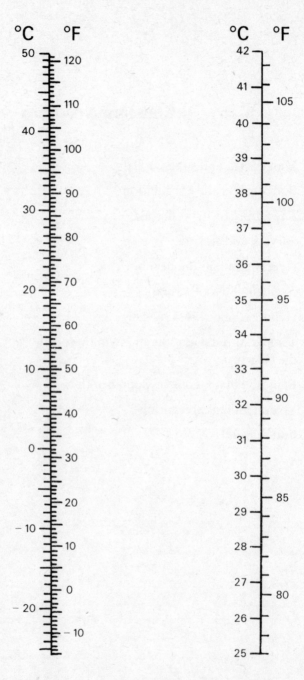

Fahrenheit/Centigrade temperature conversion chart

1

Man cooled and supercooled

The term 'hypothermia' is not of recent origin as is often supposed. It was introduced about a century ago in its Anglicized form, hypothermy, though it is not until the last few decades that hypothermia has come into general use as the condition became more clearly recognized. Hypothermia occurs when body temperature falls below normal, and this can result either from natural causes or from artificially lowering the body temperature. The body temperature referred to is the deep or 'core' temperature, which has to be distinguished from that on the body surface, or 'skin' temperature. When we use a household thermometer to take the temperature in the mouth, rectum, or under the arm (axilla), it is the core temperature we are aiming to measure. Strictly speaking, any deep-body temperature below the normal resting range of body temperature around 37 °C can be regarded as hypothermia, but in its modern clinical sense an hypothermic patient, as defined by a Royal College of Physicians Committee on Accidental Hypothermia in 1966, is one in whom the deep-body, e.g. rectal, temperature is below 35 °C.

A low skin temperature, however, does not necessarily signify hypothermia. Skin temperature is measured at the interface between the deep tissues of the body and the external environment. Its value is much more variable than core temperature, usually lying between core and environmental temperature, but it can be as low as 20 °C or less, especially on the hands, face, or feet, without there being hypothermia. Skin temperatures can be measured by thermoelectric devices, e.g. thermocouples, placed at points on the skin, but to estimate overall surface temperature special techniques such as infra-red thermography are required.

A brief history of man's experience with cold will give a perspective on how he has learnt to survive in the cold, how he has come to understand the nature of cold and its effect on himself, and how he has eventually learnt to use it for his own purposes. As we shall see, there are many examples of natural disasters caused by cold which reveal his vulnerability to hypothermia.

From prehistory, man has fought to survive in cold conditions. The glaciation periods can possibly be considered among the greatest ordeals that he has ever had to face. Palaeontological evidence suggests that hominids began in the early Pleistocene age in tropical surroundings and that they then moved into cooler Northern latitudes during the interglacial or interstadial periods when conditions during the Ice Age ameliorated. It was only later, at the beginning of the fourth glaciation some 70 000 years ago, that man appears to have remained as an inhabitant of the cool periglacial regions. Fire was clearly important to his survival in the cold. Having learned to make fire he developed the ability to keep it going, and we can surmise that the institution of gathering round the community fire helped man develop a social behaviour and transmit ideas such as how to protect himself from cold by building shelter and making clothes from animals' skins. The acquisition of clothing to improve body insulation is a prime example of the way in which human beings, by the use of intellect, have been able to adapt to cold surroundings and to extend the range of climates in which it is possible to survive.

Prolonged exposure to cold at first causes the reversible changes in physiological function which we know as acclimatization, but in the long-term there may be a more permanent adaptation by the natural selection of beneficial changes in body form and function. The possibility that climate and habitat may influence human physical form has for long been appreciated. Hippocratic medicine in the fifth century B.C. attempted to explain observed physical differences between different ethnic groups in terms of climate, and so subsequently did many more scholars. In the nineteenth century, Stanhope Smith wrote of the postulated effects of cold climate on physiognomy: '. . . severe cold has the effect of contracting the eyebrows and eyelids, closing the mouth and raising the cheeks, and this has produced the short broad face of the Tartars.' In addition to these observations on man, measurements on a large number of warm-blooded animal species led to the formulation of the Bergmann and Allen ecological 'rules', which state that the body size of a subspecies increases with decreasing temperature of the habitat, and that the relative size of protruding organs (ears, tail, limbs) increases with increasing temperature of the habitat. The general principle that body surface area relative to body weight tends to increase and that limbs may be more elongated in warm climatic regions does seem to be applicable to humans as well as animals. Evidence of man's natural acclimatization to cold or heat in different regions of the world is in fact quite difficult to find, for by behavioural responses in temperature extremes such as wearing more or less clothing, he is usually able to avoid the effects of environmental temperature stress. Even the Eskimo, who is constantly submitted to a cold climate, shows little acclimatization of the body responses as a whole. In comparison with some warm-blooded animals which have developed physiological mechanisms for lowering body temperature and hibernating in the cold seasons, man possesses a comparatively poor ability to adapt to cold.

The Greek philosophers advanced man's knowledge of cold by, for the first time, assuming that it was a property of matter and raising questions as to its nature. The injurious effects of severe cold have always been a matter for concern, as judged by some of their earlier writings. Aurelius Cornelius Celsus in *De re medicina* (A.D. 25), described cold as: 'Hurtful to an old or slender man, to a wound, to the pre-cordia, intestines, bladder, ears, hips, shoulders, private parts, teeth, bones, nerves, womb and brain. It also renders the surface of the skin pale, dry, hard and black. From this proceed shudderings and tremors.' But cold, we learn, was also deliberately used by the ancients as therapy for various ailments, particularly in the application of local cold to infected or ulcerated wounds. Galen, during the second century A.D., wrote at length on medicines which bring about a refrigerant action to the whole body. His erudition gives an interesting perspective to modern medicine: Laborit of France is given credit for the introduction in 1951 of clinical hypothermia by means of 'lytic' drugs. The cooling blanket considered a few years ago by surgeons and anaesthetists as a useful method for inducing hypothermia for operative procedures (see also p. 9) is similar in principle to that used by Sanctoria Sanctorius in c. 1660 (Plate 1), which was basically a waterproof leather bag which he called a 'balneatorium'.

Body cooling is nowadays considered as a natural recourse for patients with high fevers, but this has not always been so. The celebrated physician Thomas Sydenham broke with medical tradition in the seventeenth century by freeing his fever patients from their brood-beds designed to sweat the fever out of them and adopted instead a cooling regimen, covering the patient with lightweight sheets saturated with cold water, and allowing access of fresh air. At that time, 'temper', used as the word 'temperature' is today, described the hotness of a person and drugs were given numbers according to the heat or cold they induced in a patient.

Many more reports began to appear in the eighteenth century of the effects of cold acting as a remedy rather than as the destructive force it had previously been held to be. In 1784, Gardiner wrote of the effects of cold on a dozen children belonging to a regiment marching from Glasgow to the Scottish Highlands during a particularly severe winter. Huddled in open panniers, the children were without sufficient clothing and exposed to bitter cold. To the great alarm of the regimental physician, the children contracted smallpox along the march and in the knowledge of the then high mortality rate associated with the disease, hope for their survival was almost nil. However, to everyone's astonishment the children sustained merely a mild illness and all survived, a fact attributed, whether rightly or not, to the protection of the cold. The French physician Phillipe Pinel recorded a case of a lunatic who escaped from an asylum in the Pyrenees and was observed to run unclothed into the forest and romp and play in the snow without apparent ill-effects and became 'gradually dispossessed of his mania'. Today we would have little hesitation in ascribing the outcome of this episode simply to the effects of exhaustion, but two centuries ago it suggested the natural power of cold as a remedy.

Cold has always been a powerful enemy of both sides in military campaigns. Yet Thomas Beddoes was persuaded of the favourable effects of cold exposure upon wounded English troops retreating in Holland during the severe winter of 1794. He was convinced that exposure to extreme cold had much to do with the survival of many feverish and wounded men. Confirmation of this view came from the vivid accounts of Napoleon's Russian campaign given by Barron Larrey, Surgeon-in-Chief to the Grand Army. He told of the improved welfare of those ill with fever who were forced to seek shelter in the deep snow along the road and described the use of snow and ice in reducing pain and haemorrhage during amputations performed on soldiers in the field. Larrey reported that cold reduced the sensibility of damaged tissues and that it had a sedative effect which acted, most importantly, on the brain and nervous system. Nonetheless there is no doubt that Napoleon's army in Russia suffered huge casualties due to bitter winter conditions. Cases of freezing *en masse* are also known to have occurred in the expeditionary forces at the siege of Sebastopol in the Crimea, and in the First World War in 1914, 100 000 Turkish soldiers are reported to have died of cold in sub-zero temperatures in one region of the Turkish mountains.

Cold-climate therapy became an accepted and popular form of treatment for tuberculosis in the nineteenth century. Cold combined with high altitude was the basis of the treatment established in many sanatoria in mountain resorts. Nowhere is this better described than in Thomas Mann's great classic *The magic mountain* with its poignant account of patients undergoing the 'cure' in a sanatorium in the Swiss Alps. As a matter of fact there is little evidence to show that climate plays much direct part at all in the development and treatment of tuberculosis, though a pleasant climatic environment will have a favourable psychological effect and may help recovery. Today's powerful antituberculous drugs provide successful treatment in all climates throughout the world.

An early account of exposure hypothermia given by the Swedish Academy of Sciences in 1757 describes the case of a drunken peasant who, apparently, after being bowled over by the wind, fell asleep in the snow. He was discovered next morning frozen stiff and was put into a coffin ready for burial. Later that day a physician arrived unexpectedly and examined the body. The face and extremities were ice cold, the joints immovable, the eyes fixed open and there were no signs of breathing or heartbeat. The physician, however, thought that he could detect some warmth in the abdomen and ordered the limbs to be massaged while hot fomentations were applied to the trunk. (In fact, massaging the limbs of an hypothermic patient is not a recommended procedure in the initial stages of rewarming for it tends to bring cold peripheral blood into the central circulation and causes deep-body temperature to fall even more.) Nevertheless, as a result of rewarming the man gradually revived and was said next day to be recovering.

There are other examples of recovery after exposure to intense cold, and this general sequence of events — intoxication, unconsciousness, and exposure to cold leading to hypothermia — is familiar to hospital casualty officers throughout the world today. Let there be no doubt, however, that such spectacular resuscitation from profound cooling is not commonplace, for hypothermia is a serious condition with a high mortality unless intensive treatment is applied quickly and carefully. Emphasis is too often placed on the idea of suspended animation rather than on the inherent dangers in becoming profoundly hypothermic and during the process of rewarming.

Equally dramatic is the history of another, and usually more insidious, form of accidental hypothermia, that caused by immersion in cold water. James Currie in the eighteenth Century recorded many early cases of immersion hypothermia and he was particularly interested in the behaviour of members of crews during rescue from shipwreck. After survivors had clung to wreckage for several hours, he observed that some of them would become confused and sometimes hallucinated and would then drift away or deliberately swim away from life-rafts. Currie realized that cold had caused this behaviour, which frequently resulted in drowning, but it was not until more than a century afterwards with the sinking of the *Titanic* that his observations, long forgotten, were found to have an important meaning for maritime disasters.

The *Titanic* struck an iceberg and sank in water of a temperature of approximately 0 °C. Every one of the 1489 out of 2201 people on board who were in the water two hours after the sinking were found by the rescue crews to be dead. It was hardly likely that inability to keep afloat caused all of these deaths, for there were ample life-belts on board, yet at least one official report at the time recorded drowning as the single cause of death in every case. Some years later another passenger ship, the *Lakonia*, caught fire near Madeira and the passengers abandoned ship, this time in warmer sub-tropical waters at 17–18 °C. Within three hours of entering the water many were found dead, floating in their life-jackets. Drowning was probably the ultimate cause of death since the life-jackets were not, unlike modern life-jackets, designed to keep

7

the mouth and nose out of the water. It was clear, however, that the victims had first become unconscious because of immersion hypothermia.

Considerable experience was gained during the Second World War with hypothermia occurring after shipwrecks and of the treatment of survivors. Studies of another kind were deliberately undertaken in barbaric experiments on prisoners at Dachau concentration camp which included immersing some of them in baths of ice water until their body temperature dropped to lethal levels (below 26 °C). It was claimed that these experiments were initiated in order to study methods of resuscitation of pilots who had 'ditched' from their aircraft. Evidence given at the War Crimes Commission trial made the point that in a national emergency the performance of voluntary experiments are justified in matters of medico-military importance. However important the investigations of immersion hypothermia were at the time, it is quite certain that the experiments at Dachau were in no sense voluntary.

The modern use of hypothermia as a therapeutic procedure in clinical medicine can be said to have originated from the work of Temple Fay and his colleagues in Philadelphia in 1938–9. They took the bold step of cooling patients with inoperable cancer in the hope that lowered body temperature would affect the cancer more than the rest of the body. Deep-body temperature was artifically reduced to about 30 °C; it was found dangerous to lower it below 27 °C because of the effect on the heart's action. Unfortunately, though the procedure was at first claimed to prolong survival and to reduce pain it did not prove to be curative.

After the Second World War, W.G. Bigelow began work in Toronto with another clinical objective in mind — that of using hypothermia as a surgical tool. The advantage of deliberately reducing body temperature before surgery is that the body's rate of heat production (metabolism) is slowed, with the result that there is a fall in the oxygen requirement of the tissues. Consequently, interruption of the blood-supply (and therefore oxygen supply) by temporarily stopping the heart is less deleterious to vital organs and tissues when the temperature of the body is low. This was found to be of great therapeutic significance in heart surgery, for it opened the way to performing operations on the heart under direct vision and with the heart stopped for longer periods than had hitherto been possible. The period of safe circulatory arrest and depth of hypothermia required during such operations became the subject of much debate and research at the time. At a deep-body temperature of 28 °C, the period of circulatory arrest regarded as safe was 10 minutes. Below this temperature, ventricular fibrillation, a dangerous irregularity in the heart's rhythm which leads to complete stoppage of the heart, was found to be common.

For the purpose of surgical operation several different methods were devised to induce hypothermia. At first, the external surface of the body was cooled by ice bags, refrigerated blankets, or immersion in ice water. The immersion procedure certainly caused major transformations in the operating theatre (Plate 2). As one eminent surgeon of the day remarked, 'It is both aesthetically and surgically unattractive to see a large bath of water and ice brought along-side the patient who, with intravenous drips, intra-arterial needle, ECG leads and anaesthetizing apparatus attached, is then immersed.' A serious technical disadvantage of the sur-face cooling method was the length of time necessary to cool the patient, and to warm him again. There were also practical difficulties in management, for example in the prevention of shivering and local cold injury, and the serious problem for the anaesthetist of dealing with irregularities in the heart's action. In the 1950s, Delorme of Edinburgh and C.E. Drew and his colleagues in London took the next logical step which was to introduce blood-stream cooling. This involved circulating the blood outside the body through a pump and cooling coils and back to the body again, thereby bypassing the heart and cooling the body. Hypothermia with deep-body

9

temperature reduced to about 28 °C allowed a 10 minute safe-period of circulatory arrest, but most surgeons came to prefer more leisurely conditions provided by other methods such as the substitution of the heart and lungs by an external pump-oxygenator (cardiopulmonary bypass), which allowed up to an hour of cardiac arrest for open heart surgery. It is usual to use some degree of cooling to hypothermia in cardiac bypass operations these days; it increases the time available when temporary cessation of the circulation is required and it provides an additional safety factor in heart operations. The last three decades have witnessed remarkable developments in cardiac surgical techniques, and it can be rightly claimed that much of the impetus for these stemmed from the early work on induced hypothermia which first made open-heart operations possible.

It is important to distinguish between the effects of hypothermia, which result from lowering the temperature of the whole body, and regional cooling, which affects only a part of the body. Regional cooling can be artificially induced or it can be caused by cold injury such as frostbite. In cases of exposure to severe cold, hypothermia and regional cold injury are frequently found together. For a long time local cold has been used for treating trauma and inflammation or for inducing regional anaesthesia, and we can find references to this in Egyptian papyri going back to 3500 B.C. The local freezing of tissues in treatment (cryotherapy) and in surgery (cryosurgery) have also become an accepted part of modern medicine. The local application of a freezing substance such as liquid nitrogen to the skin for the removal or destruction of tumours, and for other surgical purposes, has been developed into a highly successful technique.

The prospect of supercooling the whole body beyond the limits of profound hypothermia for the purpose of preservation and eventual revival after rewarming has for centuries fascinated biologists, writers, and laymen. Freezing the whole body, however, is an undertaking of an entirely different

Man cooled and supercooled

order to that of regional cooling for inducing local anaesthesia. Warm-blooded animals, including man, cannot survive prolonged freezing of their entire body, at least from our present understanding, though some tissues such as blood and skin can be frozen and then restored to normal function after thawing. Semen can also be preserved in this way, a fact which has proved of enormous importance in cattle breeding. Human foetal cells placed in liquid nitrogen have been successfully preserved for as long as 20 years and are then still capable of cell division. This probably represents the longest period of time during which viable normal human cells have been arrested at sub-zero temperatures. Even more remarkable is the attempt to freeze human embryos with the eventual aim of reimplantation. When we consider the legal and ethical issues of carrying out these procedures it is clear that we are hardly able to keep pace with technological advances promised by cryogenics in medicine.

Today we are still confronted with emergencies caused by cold: shipwrecks, and accidents on mountains, moors, and in blizzards. The spectrum of potential hazards due to cold is widening as man extends his experience in underwater exploration, exploitation of the polar regions, and in space. We are also now more aware of the problem of urban hypothermia in the elderly living in cold accommodation and of hypothermia in the new-born. These problems are, of course, not unique to the present generation but, as with many other medical conditions, improvement in the diagnosis brings an increase in the number of reported cases and an apparent increase in incidence. But, at the same time, there is a greater awareness of situations where hypothermia can occur and we can take steps to guard against it. The overall incidence of cases of hypothermia in the world is probably not much greater today than it ever was, though as will be seen in the following chapters there have been considerable advances in the methods for detection, prevention, and management of the condition.

11

2

Body-temperature regulation

In winter when the energy received from the sun is reduced, many living organisms virtually cease all activity. Some molluscs retire into their shell, insects enclose themselves in their own protective covering, and deciduous perennial plants cast off their respiratory organs by shedding their leaves and become dormant. Plants have another particularly good method of surviving cold: they get rid of excess water so that the cells are less likely to be damaged by freezing. Many 'cold-blooded' animals such as reptiles and fish are at the mercy of cold conditions and their body temperature fluctuates in accord with the climate: they are 'poikilotherms'. Their body temperature drops in the cold and they become sluggish and unresponsive.

The higher animals have developed a system of keeping their bodies warm in winter and consequently are able to remain active during the cold season: they are 'warm-blooded' animals or 'homeotherms'. Not all warm-blooded animals remain active in the cold. Some are hibernators and they retain the capability of returning to the cold-blooded state, dropping their body temperature and sleeping during the winter. Some creatures, primarily birds, do not wait to combat cold weather when it comes but migrate long distances to warmer climes and have evolved methods of orientation which enable them to do so. Another method of coping with cold is possessed by animals who are able to adapt to winter conditions by a seasonal increase in the amount of fat laid down under the skin and by the growth of a thicker fur. Man does not hibernate, lay down a thick layer of subcutaneous fat, or grow fur to protect himself from cold stress in the winter. But he has one outstanding advantage: the intellectual ability to create a whole range of artificial climates for himself in order to avoid climatic extremes.

TEMPERATURE REGULATION IN MAN

Some of the coldest outdoor temperatures in regions inhabited by human beings have been recorded in the village of Verkhoyansk in Siberia where the average air temperature one January was 52 °C below zero (–62 °F). In contrast, temperatures of 48 °C above zero (120 °F) are commonly experienced by nomads living in desert and semi-desert areas of north and central Africa. Evidently the human species has a great capacity for dealing with climatic stress. As with all homeotherms, this is made possible by the thermoregulatory system of the body which acts to control internal temperature within a very narrow range.

The thermoregulatory system senses cold by means of specialized nerve endings in the skin and reacts by causing constriction ('vasoconstriction') of the blood vessels so that the insulation of the shell of the body is increased by reducing the amount of warm blood passing to the skin. This is rather like increasing the lagging round a hot-water tank in order to conserve the heat. If the primary skin vasoconstrictor defence reaction is not adequate to prevent body temperature falling then another reaction occurs, shivering, which helps generate more heat within the body. The analogy here is the stoking up of the boiler fire which causes a temporary boost in the water temperature. Conversely, in a hot climate, the temperature-regulating system at first dilates the blood vessels in the skin, which helps to dissipate heat from the body, and this is later reinforced by sweating which further increases heat loss.

The temperature-regulating system of the body is clearly very effective if it can maintain a constant body temperature in the face of extreme polar or tropical climates. Even so, temperature extremes can be tolerated for only a limited period unless there is some protection provided by shelter together with adequate clothing in the cold and fluid replacement in the heat. What is most important for survival in these environments is the behavioural response to temperature, intuitive to the native: the Eskimo wears protective clothing made of skins and fur; the tropical dweller does not venture out in the heat of the day. In this manner man avoids the effects of extreme climates without recourse to excessive shivering or sweating.

The control system

In order to understand how hypothermia develops in a species whose temperature is apparently so well regulated, it is important to consider how the control system functions. The thermoregulatory mechanism in man can be described in cybernetic terms as a loop system which is under feedback control (Fig. 1). This means that when a climatic disturbance affects the body's temperature, information about the change in temperature is fed back via the nervous system to a centre in the brain. Here the signals from the different temperature receptors are integrated and automatic control actions put in train. In response to cold these control actions are vasomotor (constricting the blood vessels in the skin), shivering, endocrine (longer-lasting chemical reactions in the body produced by hormones), and behavioural. Depending on the strength of the cold signals, the centre stimulates the appropriate control actions in order to restore body temperature back to normal once again. It can readily be seen why the temperature-regulating centre in the brain has been compared to a thermostat. In many ways this is a useful analogy though investigations have shown that its organization and function are much more sophisticated.

The control system illustrated in Fig. 1 is a simplification of the system in the body. For one thing, it is unlikely that the control loop operates in response to only one body tempera-

Body-temperature regulation

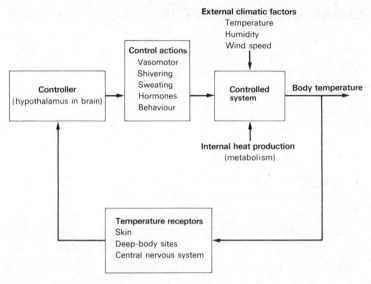

Fig. 1 Diagram of the thermoregulatory control system

ture. Many different temperature signals are transmitted to the control centre: skin temperatures which can differ widely, core temperatures which vary much less, and the temperature of the central nervous system itself. The 'body temperature' that is controlled is probably a combination of these. The whole operation of the system is, however, geared to maintain internal body temperatures within a narrow band close to 37 °C.

The precise control over body temperature outlined above is obviously directed towards maintaining temperature equilibrium, and for a number of very good reasons. If body temperature is allowed to fall too low, body functions become sluggish and eventually stop. Similarly, if the temperature rises to very high levels, body functions again become disorganized and fail. Human beings operate most efficiently at a deep-body temperature of 37 °C and most of the internal biochemical processes in the body work at their optimum at this temperature.

HEAT AND TEMPERATURE

Body temperature is an index of what is happening to the heat exchanges between the body and the external environment, and it is important therefore to appreciate the difference between heat and temperature. From early times, philosophers speculated on the nature of heat and cold and opinions had been divided as to whether heat was a substance or an effect of motion. In the eighteenth century, physicists and physical chemists came to the conclusion that what gave our senses the impression of heat or cold was the speed of motion of the constituent molecules within the body or object. This is now the accepted view and heat is regarded as a form of energy (and cold the lack of it) which is interchangeable with other forms of energy, e.g. electrical, mechanical. Heat is defined in units of energy, i.e., calories or joules (1 calorie = 4.184 joules). In conforming to the International System of Units, heat exchanges are now expressed in units of power (energy per unit time) as joules per second, or watts. The watt is a familiar unit with which we measure the power consumption of an electric fire or an electric light bulb, and it is also the unit with which we can measure the heat exchanges between the body and the environment.

A given quantity of heat energy can be distributed over a larger or a smaller piece of the same material. The average movement of the constituent molecules or atoms in the larger piece is therefore less than in the smaller. A more familiar way of expressing this is that the temperature of the larger piece is lower and the index we use is degrees Centigrade or degrees Fahrenheit, etc. Thus the quantity of heat energy in a body is measured in watts and the average movement of the constituent molecules at any one time is its temperature (in degrees).

HEAT BALANCE

Heat is exchanged between an object and its environment

when the temperature of the object and its surroundings are different. The exchange takes place across the interface (the 'skin') according to clearly defined physical laws which govern the movement of heat by the processes of convection, conduction, radiation, and evaporation. The same physical laws apply to heat exchanges between the human body and the external environment. There are, however, some important differences. In the first place there is a continuous production of heat within the body due to chemical processes which burn up oxygen and create heat (metabolism). Secondly, the surface layer of the body is not a passive structure. Its physical state can alter, for example by a change of skin blood-flow or by sweating, changes which directly affect the amount of heat exchanged through the skin.

If the core temperature of the body is to be maintained in equilibrium, then the amount of heat produced internally by metabolism must be exactly offset by the amount of heat lost by convection, conduction, radiation, and evaporation from the body surface. Heat balance is achieved when the heat produced is equal to heat lost. If a balance is not reached, as for example when the environment is so cold that the heat loss completely overwhelms the heat produced, there is a net loss of heat and body temperature drops. All the factors involved in maintaining temperature equilibrium can be accurately measured and put into the form of a heat-balance equation:

$$M = E \pm C \pm R \pm c \pm S,$$

where M is the metabolic heat production of the body, E the amount of heat lost by evaporation of water from the body, C and R the heat gained or lost by convection and radiation, c the amount of heat gained or lost by conduction, and S the amount of heat gained or lost by the deep tissues of the body. If the body maintains thermal equilibrium then the value of S (heat storage) is zero.

Several factors of the physical environment such as air temperature, humidity, radiant heat, and wind speed, combine to

17

influence body heat gains and heat losses. Air temperature affects convective heat exchanges, humidity affects evaporative heat loss, wind speed influences both convection and evaporation, and the temperature of the solid surroundings influences heat exchanged by radiation. Air temperature is most simply measured by suspending a mercury-in-glass thermometer in the air and if there is a radiant heat component in the environment the thermometer should be shielded by a metal tube. Environmental humidity is assessed by the wet-bulb temperature which consists of a thin cotton or muslin sleeve, wetted with distilled water, placed over the bulb of the thermometer. In a hygrometer, the dry- and wet-bulb thermometers are housed together in an instrument which has some mechanism for drawing air over the two bulbs. Radiant heat can be measured by a thermometer mounted inside a hollow sphere which is painted matt black on the outside to absorb radiant heat. Wind speed is measured by a vane anemometer when air movement is uni-directional and by a silvered Kata thermometer when air movement is multi-directional. With the help of these instruments, the amount of heat exchanged between the body surface and the environment can be calculated quite precisely.

Metabolic heat production (M)

There is a peat fire in North Yorkshire that is said to have been tended continuously and to have been kept burning for a hundred years. Metabolism is rather like this, continuously producing heat throughout life, the sum of all the heat-producing chemical reactions in the body fuelled by food and the intake of oxygen. Even during complete rest we continue to consume oxygen and to produce heat at the rate of about 70–80 watts (depending on the size of the person). Most people have at least a 10-fold range of heat production above the resting state which can be attained by the hardest physical work of which they are capable. The resting metabolic rate is highest in babies in proportion to their relative size. It declines gradually with age and it is approximately 10 per cent lower in women than in men (see p. 93). Metabolic heat production is largely determined by muscular activity: increasing physical work burns up more of the body's fuel and increases heat output. Internal heat production is also increased significantly by shivering. During bursts of intense shivering, the heat produced in the body may increase from a basal level of 70–80 watts up to about 400 watts. In addition, a small increase in metabolism follows the taking of a meal. This is known as the thermic response or specific dynamic action of food. The increase usually is only about 10 per cent above the basal level of heat production, with a peak occurring one or two hours after a meal, but it lasts a relatively long time (four to six hours). During the whole waking period each day the intake of food at intervals of four to six hours contributes to internal heat production. It emphasizes the importance of meals in cold conditions where the extra heat loss from the body needs to be balanced by more internal heat.

It is an eloquent comment on the effectiveness of the thermoregulatory system that if none of the metabolic heat at rest were allowed to disperse through the skin the temperature of all the body tissues would rise by 1 °C per hour (1.8 °F/h) and lethal heat-stroke levels reached in four hours.

Evaporative heat loss (E)

It is a law of physics that heat is required to turn water into water vapour. The body loses water by evaporation from the skin all the time and therefore heat is constantly being extracted. Water also evaporates from the lungs and respiratory passages. Evaporation from the skin and lungs is unavoidable and it continues without control from the thermoregulatory system. It is described as insensible water loss. At rest in a comfortable temperature approximately 30 grams of water per hour evaporates, equivalent to just over 0.7 litre (1 pint) a day. More than 70 kilojoules per hour (20 watts) of heat are extracted, which is about one quarter of the basal level of heat production by the body.

In hot climates, evaporation from the skin becomes a much more dominant and important feature, as sweating starts under the control of the thermoregulatory system. It produces heat losses of quite a different order to those of insensible water loss described above. A sweat rate of 1 litre (1¾ pints) per hour will eliminate heat from the body at the rate of 2430 kilojoules per hour (675 watts), provided all the sweat evaporates.

Heat exchange by convection (C)

If the air temperature is lower than skin temperature, the body loses heat by convection. On the other hand, air temperatures higher than skin temperature lead to a heat gain by convection. This exchange of heat occurs because as air in contact with a hot body is warmed it becomes lighter and so rises, to be replaced by cooler air. Thus it can be seen that the transfer of heat from the body surface to a colder air environment by convection will depend on the difference in temperature between the skin and the environment and on the existing air movement. If air movement over the body is increased, either artificially by an electric fan or by exposure to wind, or by making the body move in air, heat loss is further increased by 'forced convection'. Posture has a bearing on convective heat exchange. Most people bend their head forward when walking into a cold wind. This is a protective stance which minimizes the sensation of cold wind against the face but also reduces the area of the face exposed to forced convective heat loss.

Body-temperature regulation

Heat exchange by radiation (R)

Radiant heat takes the form of electromagnetic waves, of which infra-red radiation is a part. Transfer of radiant heat from one surface to another depends on the difference in temperature of the surfaces but not on the air temperature or air movement between the surfaces. As radiant heat from the sun crosses many thousands of miles in the vacuum of space before it is received by the earth, it is obvious that no medium such as air is required to transmit it. It is the character of the surface which is important in determining the amount of radiant heat absorbed or emitted. A matt black surface emits and receives radiant heat best of all but a highly polished one will also reflect it in the same way that a mirror reflects light. Human skin acts as a 'black body', which means that heat is both readily absorbed from a radiant heat source and emitted from the skin to objects at a lower temperature. Out of doors, the amount of radiant heat from the sun received by a person depends on the surface area 'seen' by the sun and this will vary with the person's posture and the sun's altitude. Normally, radiant heat will be lost from the body, clothed or unclothed, when the surroundings are at a lower temperature but there is always some solar gain, particularly when the sun is visible.

Heat exchange by conduction (c)

The amount of heat exchanged by conduction is comparatively small in normal living or working conditions because it is necessary for the body to be in direct contact with an object which is a heat conductor. The surrounding air is a very poor conductor. Heat gains can occur when, for instance, hot objects are grasped and conductive losses take place when the unshod feet are in contact with the ground. On the whole, conduction normally plays an insignificant part in heat balance but if someone lay stretched out on a cold, conducting surface much heat would be lost by conduction.

Hypothermia: the facts

Body size and shape

The rate of heat exchange is influenced to some extent by body size and shape. The loss or gain of heat by convection and radiation and the loss by evaporation is greater for someone with a larger surface area of the skin. Overall, a larger and heavier person has a bigger surface area from which to lose heat, but on a weight-for-weight basis the surface area of a light person is greater. This is advantageous to the small person in hot climates where a relatively greater rate of heat loss can be achieved but it is the heavier person who has the advantage in the cold. For a given body weight a linear body shape loses relatively more heat than a spherical shape. A possible evolutionary expression of these differences is that races of people tend to have a heavier stocky build in colder regions of the earth and a slimmer and smaller physique in the tropics.

Body shape affects heat loss in another way. Both convective and evaporative heat loss increase rapidly as the diameter of the limbs is reduced below about 10 centimetres. The Nilotes, who live in a very hot part of the Sudan, are remarkable for the linearity of their build and long attenuated limbs — a body shape that promotes body heat loss. Such a linear build may be attained by delayed maturation of the skeleton during growth in a hot climate.

ADAPTATION TO COLD

It has been said that man is a 'tropical' animal. A contention based on evidence that man originally evolved in central Africa in a hot climate. He is certainly well adapted for life in the tropics in that he has a highly effective sweating mechanism and the ability to acclimatize to work in the heat by improving his heat-regulation responses. It is much more difficult, however, to demonstrate that man can become acclimatized to cold. What we would look for as evidence of this are improved heat-producing reactions in the body and physiological changes such as in the ability to shiver or to vasoconstrict the skin blood vessels. Such changes have been reported but the evidence is equivocal and certainly not as clear-cut as those accompanying heat acclimatization.

In many animals, especially rodents, an important means of adapting to cold climates is provided by a special tissue known as brown fat. This tissue has the attribute of producing heat when the animal is exposed to cold and it provides an additional and quite different source of heat to that produced by shivering or muscular activity. During exposure of rodents to cold temperatures over many weeks the amount of brown fat can actually be shown to increase and the heat produced from it enough to render shivering unnecessary. Brown fat also plays a crucial role in arousal from hibernation in these animals. This very effective heat-producing mechanism is important in thermoregulation in the new-born human infant (see p. 56 and Plate 3), but brown fat virtually disappears in the adult human and does not seem to be vital either in the immediate response or in adaptation to cold.

The thermal conductivity of white fat is considerably less than that of skin or muscle and, in contrast to brown fat, has a poor blood-supply. Thus a fat person will be better insulated than a thin one, but there is no convincing evidence that this means of cold defence is developed physiologically by individuals habitually exposed to cold.

The various Indian tribes of Tierra del Fuego described by Darwin in *The Voyage of H.M.S. Beagle* (1890) lived with little clothing and in the most primitive huts in spite of severely cold weather. Darwin gives an account of a woman who he saw swimming during a snowstorm, with her infant clinging to her back. These people clearly demonstrate that human beings can be adapted to cold. A recent study of these same people showed them to be capable of maintaining a high metabolic heat production throughout the night without shivering intensely. Australian aborigines also appear to sleep through cold nights lying between fires, unclothed, and with little shivering. Unacclimatized white subjects under the same conditions only managed to sleep fitfully, shivering violently between periods of sleep. The aborigines appear to undergo considerable cooling of the extremities before hypothermia wakens them. They may then sit up, attend to the fire, and go back to sleep again.

These and other studies on primitive peoples who have lived for generations experiencing cold exposure with little in the way of protection suggest that adaptation to cold may develop in the long-term. For shorter periods of cold exposure, over weeks or months, there is less evidence of acclimatization except in a much more localized way in the extremities. When the hands are placed in ice water the blood vessels in the skin constrict intensely and in about a minute the fingers become increasingly painful. After a few more minutes there is usually a sudden feeling of warmth and relief from pain which is due to dilatation of blood vessels, and if the hand is kept longer in ice-water the vessels will gradually constrict again and the cycle will be repeated. This phenomenon of cold-induced vasodilatation clearly has a protective function, and what is more, it can be shown to be enhanced by continuous exposure of the extremities to cold. People whose hands are regularly placed in cold water, such as fishermen, show a local adaptation, the initial constriction is less severe and the dilatation occurs more rapidly and lasts longer.

INSULATION OF THE BODY FROM COLD

Clothing

The major protection most of us receive in cold conditions is not from our own bodily defence mechanisms but from the extra insulation achieved by clothing. The insulation provided by clothing is due to the air trapped between different layers and the air trapped between fibres or hair of the clothing material. Cotton wool and down feathers are equally good at trapping air but whereas cotton wool becomes easily compressed, down feathers quickly recover their shape after compression and retain their insulation. Similarly, fur coats providing the best insulation are composed of strong hairs which spring back into shape after compression.

Although the trapping of air in the interspaces of clothes provides extra insulation and prevents the loss of heat, strong wind playing on to the clothes can penetrate these spaces and reduce the insulation. Close-woven cloth can protect against this as also can impermeable wind-proof material. The big disadvantage of impermeable clothes, however, is that water evaporated from the skin is trapped inside the garment and in cold weather this water condenses on to the skin. If the outside temperature is below freezing the water inside the impermeable layer may even freeze. Condensation of water from inside and rain from outside both destroy the insulating capacity of a garment. Impermeable clothing is less likely to cause condensation problems in the cold if the wearer is not sweating.

Another way of preventing heat being lost from the body is to wear a layer of clothing that will stop heat loss by radiation. The best way to accomplish this is to surround the body with shiny metallic material which will reflect the radiant heat back to, and will not absorb heat radiated from, the body. Practical examples of this type of protective covering are found in the metallized 'survival' blanket used to cover exposure hypothermia patients during rescue, and in hospital during rewarming, and also in the 'space-coat' type of garment which has been advocated for preventing hypothermia in elderly people (see p. 125).

A unit of insulation, the *clo*, has been developed in order to standardize the insulation value of clothing. Thus 1 clo is the insulation provided by clothing sufficient to allow a person to be comfortable when sitting in still air in a room at a temperature of 21 °C. In Table 1 a number of different clothing outfits

Hypothermia: the facts

Table 1. *Insulation values of different clothing*

Clothing ensemble	Insulation (clo units)
Shoes and socks	0.07
Briefs and vest	0.1
Light sleeveless dress, cotton underwear	0.2
Tropical clothing: shorts, open-neck shirt with short sleeves, socks and sandals	0.3–0.4
Light trousers, short-sleeve shirt	0.5
Thick long dress, full-length slip	0.7
Light trousers, long-sleeve shirt, thin sweater	0.8
Light trousers, long-sleeve shirt, jacket	0.9
Light business suit, shirt, underwear	1.0
Heavy traditional European business suit, shirt, underwear	1.5
Cold/wet uniform (U.S. army): cotton/wool underwear and shirt, water repellent-wind resistant trousers and field coat with wool liner, wool socks	1.5–2.0
Heavy wool-pile polar suit	3.5
Lynx-fur suit	5.2
Red fox-fur suit	7.8
Arctic sleeping bag	up to 12.0

are listed together with their clo values. The amount of clothing needed for insulation in different climates obviously varies with the amount of physical activity. In a freezing climate of 0 °C the insulation needed for heavy work may be only 1 clo but for light or sedentary occupations 3 clo, and during sleep 7 clo. A suit of more than 3.5 clo is too heavy and cumbersome to allow much accurate movement. The maximum insulation of Arctic sleeping bags is about 12 clo.

The clo unit is the most widely used unit of clothing insulation in most parts of the world and it has been employed extensively by the armed services where protective clothing is designed for men serving or undergoing training in cold environments. Another unit, the tog, is commonly used to evaluate insulation by textile manufacturers, particularly in the U.K. It is a smaller unit than the clo (1 tog = 0.645 clo) and it is used, for example, to measure the insulative capacity of the duvet bed cover now becoming more widely used in the U.K.

Body-temperature regulation

Shelter

Housing not only helps to insulate from the cold but provides means to increase heat input to the body through indoor heating. Out of doors, simple shelters help to reduce the effects of cold by protection against wind, rain, and snow and to improve insulation. It is particularly necessary to try to reduce the cooling power of the wind, which blows away the warmed layer of air surrounding the body and penetrates the air spaces in clothes.

The Eskimos have succeeded in creating a microcosm of an environment in which to live which provides them with a semi-tropical climate in the middle of the polar zone. The traditional igloos are low and round, about 3–4 metres in diameter, and built of snow blocks. The entrance is through a long tunnel sunk into the ground or snow, and communicates with a pit inside the igloo. Ventilation in such a house is perfect. Hot air escapes through small apertures in the domed roof which can be stopped up at will. Cold air enters from below through the tunnel which often has no door. It can enter only as fast as the hot air is allowed to escape. As it rises, the cold air spreads slowly without noticeable draught. It is heated at the level of lamps burning seal fat in the pit of the igloo and is at a comfortable temperature by the time it reaches the occupants who live on a platform inside.

MEASUREMENT OF BODY TEMPERATURE

Galileo is considered to be the inventor of the thermometer in 1603, though Sanctorius used it for clinical purposes and in 1612 gave the first description of the 'baro-thermoscope' which he regarded as a 'very ancient instrument'. Some of the earliest thermometers using mercury as the thermometric fluid were made in the seventeenth century by the Academia del Cimento in Florence. Fahrenheit began work on thermometers in 1706, making thermometers which agreed

27

much more exactly than any previous instruments, and Celsius introduced the Centigrade scale in 1742. The mercury-in-glass clinical thermometer at present remains the traditional instrument, though electronic instruments are also available for measuring body temperature. In Great Britain each clinical thermometer sold for general use has been tested to comply with specifications laid down by the British Standards Institute (BSI 691:1979). Clinical thermometers of present-day manufacture are graduated in degrees Centigrade (or Celsius) but there are many clinical thermometers in everyday use which are graduated in degrees Fahrenheit. A temperature conversion chart (Appendix 1) is given at the end of this book.

In recent years, one of the most important contributions to improvement in diagnosis of hypothermia has been the addition of a 'low-reading clinical thermometer' to the doctor's bag. Most general-purpose clinical thermometers used in hospitals and at home measure body temperature in the range 35–42 °C (95–108 °F) — quite inadequate to register levels of hypothermia. Low-reading clinical thermometers are available from chemist shops and are subject to British Standards requirements. They measure body temperature in the scale range 25–40 °C (77–104 °F).

Core temperature is usually measured by a thermometer placed in the mouth (oral temperature) and this is a satisfactory method in warm indoor conditions and in hospital wards. Errors arise if there is mouth-breathing or talking during the measurement, or if hot or cold drinks or food have been taken just previously. The tissues of the mouth are readily affected by cold or hot environmental conditions so that taking an oral temperature in temperature conditions below about 15 °C can give a false low reading. The time taken for a clinical thermometer to indicate deep-body temperature depends more on the condition of the subject and on the method of using the thermometer than on the thermometer itself. A thermometer placed in a bath of water will give an accurate reading within a few seconds. In the mouth about two to four minutes will probably be sufficient for the thermometer to give a steady reading. This time lag is inevitable. It is necessary to allow temperature equilibration to take place in the mouth and it will not be shortened by using more accurate thermoelectric devices for temperature measurement.

The rectal temperature is a more reliable measurement of deep-body temperature and on average gives a temperature about 0.5 °C higher than oral temperature. Cold blood coming from chilled legs tends to lower the reading and warm blood from exercising leg muscles tends to raise it. Another reliable index of core temperature is the urine temperature, providing a sufficient volume of urine (100 ml or more) can be voided when the measurement is taken (see p 45 and Plate 5). For the purposes of research and continuous temperature monitoring, special thermoelectric devices are used which may be placed in the ear or oesophagus, and sometimes the temperature in the intestine is recorded by swallowing a radio-pill which transmits body temperature to a radio receiver outside the body (Plate 4).

The internal temperature of warm-blooded animals does not remain strictly constant during the course of a day. In man it may be 1 °C (1.8 °F) higher in the evening than in the early morning. This may be due partly to warmer environmental temperatures in the evening and to the effects of taking food but it is also due to the effects of an inherent daily rhythm.

3

The nature of hypothermia

In this chapter we shall try to answer the question 'What is hypothermia?' and to show what effect low core temperature has on body functions. As we have seen, the finely adjusted physiological control by the thermoregulatory system, the behavioural responses, insulation provided by clothing, and possible long-term cold adaptive changes in the body, together normally provide a very adequate defence against cold conditions and the threat of hypothermia. There are, however, two main factors, which we shall call internal and external, that can cause core temperature to fall. Internal causes arise from disorders of the thermoregulatory system which may be due to dysfunction of normal physiological processes of temperature regulation, from pathological changes in the body due to disease, or from the action of drugs. External causes are environmental: severe temperature conditions can overwhelm existing body defences and destroy the heat balance. These internal and external factors are additive. Thus a moderately cold environment normally easily tolerated by a healthy individual could become too cold for a sick person to contend with and may lead to a fall in body temperature.

Various terms have been used to describe different types of hypothermia. One method is to classify according to the depth to which the core temperature is lowered, without taking into account the actual cause of hypothermia. Information on the extent of body temperature lowering is very important, particularly for the subsequent treatment required in hospital. Hypothermia is accordingly classified in the following way:

The nature of hypothermia

	Deep-body temperature:
Mild hypothermia	34–35 °C
Moderate hypothermia	30–34 °C
Severe hypothermia	less than 30 °C

There are also descriptions which refer to the time-scale over which hypothermia develops. For example, spontaneous hypothermia may occur suddenly without any previous indication and for no apparent reason; chronic hypothermia is a term sometimes used to indicate sub-clinical, i.e. near 35.0 °C, levels of body temperature which may exist in some elderly, very ill, or moribund patients for long periods of time; and intermittent hypothermia describes a type of hypothermia occurring at intervals, perhaps a few times a year. These are useful descriptive terms which can provide insight into the probable cause of hypothermia.

Neither the depth of hypothermia nor the time-scale are sufficient alone to provide suitable epidemiological information about the condition, valuable though both of these dimensions may be. A simple general classification which is helpful in the diagnosis and notification of hypothermia is to divide cases into the following four categories.

Primary hypothermia

This is due to an inherent impairment or dysfunction of the thermoregulatory system itself. It may apply, for example, to the failure of thermoregulatory processes with age in some individuals or to immature development of these processes in the premature infant. Some rare cases of intermittent primary hypothermia are thought to arise from developmental anomalies in the human brain, particularly in the region of the hypothalamus.

Secondary hypothermia

Many clinical conditions can lead to intolerance to cold and in

31

these cases low body temperature is a symptom of the underlying disease rather than the primary event. Some of the disorders associated with hypothermia are acute, such as those
following a heart attack (myocardial infarction) when many
bodily functions including thermoregulation are indirectly
affected, while others may be due to long-standing illnesses
causing immobility.

The most important causes of secondary hypothermia are:

(1) lowering of the body's metabolism so that internal heat
    production is reduced. This can result from diseases of the
    endocrine system such as hypothyroidism and hypopitui-
    tarism and from the inability to utilize sugar, as in diabetes
    mellitus;

(2) immobility caused by coma or chronic diseases such as
    parkinsonism, paralysis, and severe arthritis, which
    reduces internal heat production. Muscular activity plays
    an important role in helping to generate heat and patients
    who are forced to lie immobile in bed are denied this
    defence against cold;

(3) poor circulation, especially if the circulation is in danger
    of collapse as the result of coronary thrombosis, pulmon-
    ary embolus, or overwhelming bacterial infection, will
    significantly reduce the effectiveness of thermoregulatory
    responses;

(4) nervous diseases such as stroke, brain tumour, or the
    effects of head injuries, which may impair the function of
    the thermoregulatory centre or peripheral nerves control-
    ling body temperature. With mental impairment or confu-
    sional states a patient may be unaware of the danger of
    cold conditions and be unable to take action to protect
    himself; and

(5) many drugs acting on the central nervous system can be
    implicated in causing secondary hypothermia. They not
    only directly affect temperature regulation responses but
    also lessen the subject's awareness of environmental
    hazards. Excessive amounts of sedatives, tranquillizers
    and antidepressives are often found to act as hypothermic

agents, as also is the intake of large quantities of alcohol (see Chapter 7).

Accidental hypothermia

This is probably the most familiar type of hypothermia, usually recognized from the circumstances in which it occurs. The condition arises unintentionally in the healthy person as the result of an accident or error of judgement which results in abnormal cold exposure. Many examples are described in Chapter 6, encompassing off-shore and diving accidents, exposure on mountains and during polar expeditions, and also accidents which can occur in the urban environment.

Therapeutic hypothermia

This is a type of hypothermia artificially induced for clinical purposes, a brief account of which has been given in the introductory chapter. The technique is used rather less now but it has played a very important role in the development of open heart surgery and has provided valuable information on the changes associated with other forms of naturally occurring hypothermia.

EFFECTS OF BODY COOLING

There is a range of climates in which the body remains in a thermally neutral state while the deep-body temperature is kept constant at the normal level near 37.0 °C. In this climatic zone, body temperature is stabilized by vasomotor control (constriction and dilatation of blood vessels in the skin). The thermoneutral zone can be described as one involving the least thermoregulatory adjustment. At the lower boundary of thermal neutrality there is a critical temperature which is the environmental temperature at which cold defence mechanisms are brought into play. The critical temperature will depend on whether the person is well clothed or not, active or resting, as various other components of the environment such as wind speed. If the individual is unclothed and resting, the ideal still-air ambient temperature at which he remains thermally neutral and feels comfortable is 27–29 °C (81–84 °F). Under such conditions the temperature of the core is about 37 °C and the average skin temperature about 33 °C.

If ambient temperature drops below the lower critical level the temperature difference between the skin and the environment is increased and this causes an increase in heat loss through convection and radiation. The body reacts to this by constricting the skin's blood vessels, which reduces the temperature difference between the skin and environment and counteracts the increased heat loss. There is, in fact, another mechanism which helps to reduce the amount of heat lost from the skin in cold conditions. Heat brought from the core of the body in blood flowing through arteries to the skin is 'exchanged' with the deep veins which run in close proximity to the arteries in the extremities of the body. Blood in these deep veins comes from superficial veins in the skin surface and is therefore cool. The result is a 'countercurrent exchange' so that some heat being brought to the extremities via the arteries is put back into the core via the deep veins before it even reaches the skin surface.

As body temperature starts to drop below normal, the skin vasoconstriction becomes more intense, reducing heat loss from the skin. At this stage the muscles of the body begin to tense prior to shivering. Shivering causes a considerable amount of heat to be released internally within seconds or minutes, while at the same time failing to produce any coordinated movements by the muscles. This is because the tremor occurs in both sets of muscles which normally oppose one another, i.e. when a movement is made one muscle group contracts and the opposing group relaxes. The antagonist muscles thus contract against each other and produce heat rather than movement. Any purposeful movement of the muscles or physical activity tends to inhibit shivering. If for

34

any reason vasoconstriction and shivering fail to restore body temperature towards normal, hypothermia develops as core temperature reaches 35.0 °C.

Mild hypothermia (core temperature 35–34 °C)

Below a deep-body temperature of 35.0 °C it is usual for some clinical signs of disorder to begin to appear. There is still intense vasoconstriction of the skin, pallor, and usually shivering but also there are signs of muscular weakness and incoordination. Walking may begin to be affected and the individual may stumble or fall. Consciousness is not impaired but his mental state may become perceptably dulled, often shown by a lack of response to the rigours of the environment and difficulty in understanding the situation. Elderly people in their own homes often become sluggish in their movements or sit quietly immobile in a chair if they become hypothermic.

Moderate hypothermia (core temperature 34–30 °C)

In the transitional zone between 'safe' core temperatures just below 35 °C, when physiological mechanisms for heat conservation and heat production still operate, and the 'danger' zone near 30 °C when they largely cease, shivering usually diminishes though skin vasoconstriction is still intense. The pulse-rate is now much reduced and weak and breathing becomes shallow, usually less than 10 breaths per minute. Consciousness may be lost at body temperatures between 32 °C and 30 °C and by then shivering has usually stopped and the fall in body temperature becomes more rapid.

Severe hypothermia (core temperature less than 30 °C)

The unconscious hypothermic patient has a death-like appearance, is very cold to the touch, and there is extreme pallor, often with the absence of a pulse and an associated generalized rigidity. The skin may develop blue patches and become puffy (oedematous). Breathing is barely perceptible. At body temperatures of 28 °C and below there is a danger of heart irregularity, first fibrillation (rapid and chaotic beating) of the atria (upper chambers of the heart) and then of the ventricles (lower chambers). Ventricular fibrillation is rapidly fatal unless defibrillation (electric shock treatment) can be applied to the heart and resuscitation procedures commenced. Below body temperatures of 25 °C the physiological mechanisms for heat conservation are absent and heat is lost passively. Like the poikilothermic cold-blooded animal, the hypothermic patient then becomes completely at the mercy of the environment.

CHANGES IN BODY FUNCTION DUE TO HYPOTHERMIA

Apart from external signs of the effects of cooling such as skin pallor, shivering, weakness, and dulling of mental processes, other, more general, changes take place within the body. The progressive development of hypothermia is associated with a gradual slowing down of all biological processes. This applies to metabolism, blood circulation, respiration, the functioning of the nerves and intellectual function. Some processes such as inflammation are damped down, but this affords only a temporary alleviation of symptoms, not a cure. So a mild degree of hypothermia can be induced in order temporarily to slow down a pathological reaction, to reduce pain, or to enable a patient to be moved to a more suitable place for treatment.

Energy metabolism

The body's oxygen consumption and metabolic rate gradually decline as body temperature falls if shivering, which increases oxygen consumption, is prevented. Shivering will usually have stopped in any case when body temperature is below 30 °C. When the core temperature is 32 °C, the overall oxygen consumption has fallen by about 25 per cent, at 28 °C by about 50 per cent and at 10 °C by more than 90 per cent so that there is then scarcely any metabolic activity. The rate at which different processes in the body decline during progressive hypothermia is not uniform. Active metabolic processes such as muscular work and also rhythmic processes are particularly depressed, while slow physico-chemical events such as the diffusion of substances in body fluids are less affected. Contractile processes are in an intermediate category and muscle cells in the heart, for instance, can continue to contract on being artificially stimulated at a body temperature much lower than that at which all rhythmic activity of the heart has ceased.

The heart and circulation

The heart-rate is often at first raised in a patient with mild hypothermia. This can be due to the effects of shivering or sometimes to an 'alarm' reaction, especially if the patient has been suddenly immersed in cold water. As the core temperature falls, heart-rate likewise decreases. The work that the heart muscle is able to do is affected when the tissues are severely chilled so that the output of the heart is directly reduced. Under these circumstances there would normally be a tendency for blood-pressure to decrease but in fact this only appears to happen in severe hypothermia, partly because the reduced output of the heart is compensated by intense constriction of blood vessels in the periphery tending to maintain blood pressure by increasing the resistance. The electrocardiogram (ECG) frequently shows some degree of block in the heart's conducting system and often a characteristic deflection in the ECG trace known as a 'J-wave'. In the past, the J-wave has been regarded as a quite specific indication of hypothermia, but it is not present in all hypothermic cases and there is some doubt as to whether it can be considered to be a diagnostic sign. Disturbances in the rhythm of the heart beat (arrhythmias) are common in all types of hypothermia, especially in old people. The irregularities are usually not dangerous when they involve fibrillation of the atria of the heart and they are often only transient. The gravest hazard is the onset of ventricular fibrillation, especially in patients with diseased hearts. Ventricular fibrillation rarely occurs at core temperatures above 32 °C but it may be precipitated during either cooling or warming at body temperatures below 32 °C.

The respiratory system

Breathing tends to become slower and shallower as hypothermia deepens. There may be an initial stimulation as there is in heart-rate if cooling starts rapidly with sudden cold water immersion. This is familiar as a 'gasp' reflex when water may be sucked in involuntarily, and overbreathing occurs because of the very large cold stimulus to the skin. Respiratory failure may ensue at low body temperatures. A reduction in ventilation and depression of the cough reflex quite often leads to collapse of some segments of the lung in severely hypothermic patients. This in turn increases the possibility of lung infection and pneumonia. The typically slow, sighing, almost inaudible breaths contrast markedly with the rapid or stertorous breathing of hypothermic patients in whom pneumonia has developed.

Body fluids

Cold 'diuresis' is a familiar effect of low temperature conditions: the volume of urine and the frequency of micturition increase. There are a number of explanations for this phenomenon, which is variously attributed to a decrease in the (tubular) function of the kidneys in holding back water, to an initial increase in blood-pressure in cold conditions which enhances the formation of urine, and to the shunting of blood from a cold vasoconstricted skin to the blood vessels supplying the kidney which again increases urine formation. Cold diuresis, however, is not necessarily associated with a drop in deep-body temperature. The total kidney function appears to decline when body temperature falls significantly and blood flow to the kidneys is reduced. Rarely, depression of the tubular reabsorption of water is the dominant effect in which case a large volume of urine is produced (polyuria), and sometimes there is almost complete absence of urine during hypothermia depending on the degree of kidney failure or structural damage.

The overall loss of fluid from the body is decreased in hypothermia, except in the few cases where there is polyuria. Usually urinary volume is low. Water-loss through the skin and respiratory passages is also reduced. There is an abnormal distribution of water in the fluid compartments of the body with a tendency for excess fluid to accumulate in the tissue spaces and to cause oedema (waterlogging in the tissues). The loss of fluid from the blood compartment of the body during hypothermia leads to a condition resembling 'shock' with a sluggish flow of blood in the capillaries and poor blood-supply to the tissues.

In many cases of hypothermia there is also a striking increase in blood acids ('metabolic acidosis'), especially in those whose hypothermia was preceded by exhaustion and prolonged shivering. This is caused partly by poor blood-supply to the tissues resulting in retention of carbon dioxide.

Central nervous system

There are a number of similarities between the effects of lowering body temperature and of general anaesthesia. The level of consciousness and alertness gradually decreases as body temperature drops towards 30 °C. Responses become slow and reflexes sluggish and speech becomes increasingly difficult. Consciousness is usually lost between 32 °C and 30 °C but there are occasional exceptions. At 27 °C body temperature a patient may sometimes grunt when questioned but below 26 °C he usually fails to respond to any stimulus whatever. Voluntary movements gradually become slower and a simple movement such as touching the nose normally accomplished within one second may take a patient 15–30 seconds when his body temperature is at 30 °C. Muscle rigidity is a striking feature, making it difficult to extend the limbs and sometimes there is pronounced neck stiffness. At 30–31 °C body temperature the pupils of the eyes react so slowly that the light reflex may be wrongly assumed to be absent. The pupillary light reflex, however, does disappear at lower body temperature. There is, as we might expect, considerable variation from one person to another in these responses and there is no exact temperature at which a given response will disappear on cooling or reappear on rewarming.

4

Surveys and statistics

Since the early 1960s, when hypothermia became recognized as an urban problem in the U.K. and documentation began, the statistics of hypothermia have been the subject of much controversy. To focus attention on the statistics in the U.K. immediately suggests that it is a condition encountered especially in those islands. Indeed, experience with it in the urban setting is probably greater in the U.K. than anywhere, but that is not to say that the disorder is less important or is found less frequently in other countries.

The term 'urban' is used to indicate that hypothermia occurs in indoor environments. It does not of course mean that hypothermia is any less of a problem in rural areas. There is controversy over the numbers of cases involved because it is difficult to distinguish the condition of hypothermia as a primary cause or just a symptom of disease, and there are situations when it may go unrecorded. Even more confusing are situations when false readings of low body temperature are obtained, such as can occur when mouth temperatures are taken in cold ambient conditions. Added to this, hypothermia has come to be regarded as a sensitive medical and social problem closely connected with housing standards and the economics of home heating. Thus when prevalence figures for hypothermia are used, the importance of urban hypothermia is sometimes overemphasized.

Another major difficulty in interpreting statistics should be mentioned. Accidental hypothermia is usually evident from the circumstances of its occurrence, e.g. when it occurs through accidental exposure in cold mountain terrain or in off-shore incidents. These cases are more often than not well-documented. On the other hand, no-one knows with certainty how often hypothermia occurs in association with other disorders in old people or whether a large number of elderly people are precipitated into secondary hypothermia in the winter. Mortality statistics do not help very much, for the failure to maintain a constant internal body temperature can, in one sense, be considered to be part of the process of dying and consequently it is often difficult to judge whether hypothermia is the result or cause of a fatal illness. The condition of hypothermia is silent, it leaves no trace, and there is usually no means of showing that it was present if recovery occurs. We have to reconcile the suggestion sometimes put forward that many thousands of elderly in the U.K. may die as the result of hypothermia in the winter, with statistics given by death certificates which testify to a total of only a few hundreds of deaths associated with hypothermia each year.

HOSPITAL STUDIES

An indication of the prevalence of hypothermia in a population can be obtained from a study of hospital admissions. In the U.K. in the early 1960s a series of small regional studies were carried out. An early one in Scotland described 23 cases of elderly patients admitted to hospital with hypothermia during the previous three years. The deep-body temperature on admission, measured by a rectal thermometer, ranged from 22.8 °C to 31.9 °C and only seven of these patients survived. Another 32 cases of hypothermia were admitted to a hospital in London, half of whom were seen during the very cold winter of 1962–3. Ten of these elderly patients were found lying on the floor, with profound hypothermia; the others suffered from a lesser degree of exposure: ten were in bed, two were sitting in a chair, and one developed hypothermia in hospital.

The severe winter of 1962–3 led to the formation of a Committee on Accidental Hypothermia by the Royal College of Physicians of London which decided to survey the frequency of hypothermia in all admissions to selected departments of ten hospitals in England and Scotland, extending from Exeter to Aberdeen, during a three-month period (February to April) in 1965. All patients with a mouth temperature below 35 °C had their deep-body temperature checked by a thermometer placed in the rectum for five minutes. It was found that 126 patients of those admitted to hospital during this time were suffering from hypothermia. This amounted to nearly seven patients per 1000 admitted to hospital, and of these 42 per cent were over the age of 65. Projected over England and Scotland as a whole it would mean that about 3800 elderly patients were admitted to hospital with hypothermia during the three months of the survey. During the coldest week, when the average minimum ambient temperature was −4 °C, 19 patients with hypothermia were admitted, the highest number for any one week. In the three weeks of the winter when the minimum air temperature exceeded + 4 °C, hypothermia admissions averaged only four per week. It looked, from these observations, as if the incidence of hypothermia was directly related to low environmental temperatures. However, some cases of hypothermia occurred when climatic conditions were mild, which suggested that environmental temperature, though important, was not the only causative factor. It was found that there were slightly more female than male patients in the sample and that the largest percentage of these admissions were infants under one year old and elderly over 75.

The mortality in these hypothermic patients, even with hospital treatment, was high, emphasizing once more the serious nature of the condition. In the Royal College survey, 37 per cent of the 126 hypothermic admissions died. If this is considered in terms of the body-temperature levels on admission, 73 per cent of those with severe hypothermia and temperatures below 30 °C died, as did 33 per cent of those with temperatures between 32.9 °C and 30 °C, and 32 per cent

43

between 34.9 °C and 33.0 °C. It is important to recognize that all age groups were included in these figures and that most cases were suffering from secondary hypothermia, i.e., other diseases were present. There were, however, 17 patients in whom primary hypothermia or accidental hypothermia was the principal diagnosis. Ten of these patients were over 65 years of age and half of them did not survive.

A second Royal College of Physicians survey was conducted in one London hospital group (January to April) in 1975 on the elderly admitted to hospital with hypothermia. This time it was found that 3.6 per cent of all elderly patients over the age of 65 were hypothermic, a prevalence much higher than that in the first survey 10 years before when only 1.1 per cent of elderly over 65 were found to be hypothermic (0.7 per cent of total admissions, all ages). The explanation for this difference is open to conjecture. It may have been due to differences between the populations from which the patients were drawn in 1965 and 1975, or it may have been that the hospitals in the two studies admitted a different age-spectrum of patients. Such differences in prevalence figures emphasize that caution should be exercised in drawing conclusions from studies involving relatively small numbers of patients and different population samples.

If we look at hospital admissions another way and relate the numbers to the 'total population at risk' in the community for each age range, then we obtain another perspective of the distribution of patients with hypothermia.

In the North-West Region Health Authority in the United Kingdom, the International Code 788.9 has been used to record the admissions to hospital with hypothermia. The code brackets hypothermia with other general descriptions of cold exposure including 'chills and rigors'. On this basis it was found that, during 1972-77, 1 in 2000 of the 0–4 year-old population, 1 in 100 000 of the 10–40 year-old population, and 1 in 1300 of the over-80 year-old populations at risk were admitted to hospital with either a primary or secondary diagnosis of hypothermia. Again this showed the greatest prevalence at the extremities of the population age scale.

POPULATION SURVEYS

Most of the studies on hypothermia in the community in the British Isles have been conducted with one particular objective in mind. This is to try to estimate the prevalence of hypothermia in the elderly population who, from the evidence of surveys in the early 1960s, were considered to be at greatest risk. The elderly account for no small proportion of the community, more than 8 000 000 over 65 years of age in Great Britain before the 1981 census. In 1967, two small surveys of old people living at home showed a surprisingly large proportion (between 5 and 11 per cent with mouth (oral) temperatures of 35 °C or below during the winter months). Oral temperatures are, however known, to be an unreliable measure of deep-body temperature in cold ambient temperatures, and it was only with the development of the Uritemp technique (Plate 5) by Dr. R. H. Fox and his colleagues that it became practicable to measure the deep-body (urine) temperature of a large number of old people living in their own homes. The first full-scale domiciliary investigation of indoor and deep-body temperatures was carried out in 1972 in a National Survey based on a random sample of 1020 people over the age of 65. Morning and evening measurements were made of urine, oral, hand (skin), and environmental (living room, bedroom, and outdoor) temperatures. The average winter outdoor temperatures in 1972 (7 °C in the morning and 8 °C in the afternoon) were slightly higher than normal for the time of year, but in 75 per cent of cases the living-room temperatures in the morning were at or below 18.3 °C, the minimum recommended by the 1977 Parker Morris Report on Council Housing, and in 10 per cent the room temperatures were very cold, at or below 12 °C. 0.5 per cent of the elderly population surveyed were just in the hypothermic range, with urine temperatures in the morning between 34.2 °C and 34.9 °C. None, therefore, could be considered moderately hypothermic (less than 34 °C). However, between 5 and 10 per cent of the population had low urine temperatures just above the hypothermic level between 35.5 °C and 35.1 °C in the morning. These people were thought to show some degree of thermoregulatory impairment.

The salient feature of this study was the finding that only a small proportion of old people at home were actually hypothermic but that there was a significant number of elderly people with morning low-body temperatures in the region of 35.1 °C to 35.5 °C. In drawing conclusions from these figures three important facts need to be borne in mind. First, hypothermia was only identified in *morning* urine measurements in 1 elderly person in 200. In the afternoon, *no-one* had a hypothermic deep body temperature of 35.0 °C or less and very few had afternoon temperatures between 35.5 °C and 35.1 °C. Secondly, the hypothermia detected in the mornings was of a mild degree, mostly with urine temperatures at or just below 35.0 °C, the lowest temperature being 34.2 °C. No cases of moderate or severe hypothermia were found; such cases presumably would most likely have been admitted to hospital. Lastly, because of the apparent transient nature of the hypothermia depending on the time of day, it should be emphasized that to project the survey figures to predictions of the prevalence of hypothermia in the elderly population as a whole could be misleading.

REGISTRATION OF DEATHS FROM HYPOTHERMIA

The Office of Population Censuses and Surveys routinely publish statistics based on death registrations which are classified according to internationally agreed rules. Until recently this classification did not include a category exclusive to hypothermia or accidental hypothermia. Instead, hypothermia had been classified under a variety of headings including 'chills and rigors', 'excessive cold', 'hunger, thirst, exposure and neglect', and, in the newborn, 'cold injury syndrome'. Furthermore, under the rules for assignment to a single cause of death, hypothermia was treated as a symptom and disregarded when another (usually more easily recognizable) cause such as bronchopneumonia was also recorded on the death certificate.

The number of death certificates issued in England and Wales where hypothermia was specified as the underlying cause, and the total number of deaths from all causes where there was mention of hypothermia, are given in Table 2 for the period 1971-8.

Table 2. *Hypothermia registered on death certificates (O.P.C.S. published data for England and Wales)*

Year	Deaths with hypothermia as underlying cause	Total deaths with mention of hypothermia
1971	15	420
1972	20	492
1973	21	437
1974	16	401
1975	25	511
1976	21	585
1977	15	613
1978	21	708

It is clear from this Table that hypothermia was rarely registered as the principal or underlying cause of death and that the total number of deaths thought to be directly caused by hypothermia varied little from year to year. On the other hand, hypothermia was mentioned much more frequently as a secondary cause of death. That the numbers in this second category appear to have increased during the 1970s must take account of the fact that clinical awareness of the condition had increased during the decade.

In 1979, there was a revision of the International Classification of Diseases which assigned hypothermia into different categories to those used for the previous 10 years. The most recent annual statistics for hypothermia, though roughly of the same total order as before, are therefore not strictly comparable to previous years.

The U.S. Government's National Center for Health Statistics recorded the total number of hypothermia deaths from infancy up to the age of 100 years in the U.S. as 466 in 1970, increasing to 634 in 1977. During the seven-year period the death-rate for people younger than 65 ranged from 1.0 to 1.7 deaths per million. For those older than 75, the death-rate from hypothermia was 11.5 per million in 1970 rising to 17 per million in 1977. The figures for the total number of deaths associated with hypothermia in the U.S. thus closely resemble the yearly totals for England and Wales shown in Table 2. Since the population of the United States is roughly four times greater than that of England and Wales the problem of hypothermia would appear to be much smaller in the U.S. relative to the size of the population at risk. However, clinical awareness of the condition during the 1970s was probably greater in the U.K. and this may account for some discrepancy in the frequency with which hypothermia deaths were registered. There are also considerable differences in the range of climatic temperatures in the two countries which make a simple comparison more difficult. The U.K. enjoys more equable maritime climate than the U.S.A. where in the northern American states continental and polar air masses produce sub-zero temperatures in winter, and the southern states generally remain warmer than the U.K. throughout the year.

As might be expected, the statistics from both the U.K. and the U.S.A. show that most deaths associated with hypothermia occur during the winter months. The general pattern of higher risk of hypothermia in the elderly, however, continues throughout the year and is not confined to the winter months. Cases of hypothermia in the elderly and accidental hypothermia in younger adults are evidently not reserved to cold climes nor necessarily to the coldest months.

Surveys and statistics

SEASONAL MORTALITY

Since 1841 when reliable population statistics in England and Wales first became available, there has been a gradual fall in yearly mortality rates. After about 1930, total (crude) death-rates have actually stopped falling and levelled out, but if the numbers are standardized to take account of the trend to increasing average age of the population then mortality rates have decreased steadily during the period 1841 to 1980.

In many countries with a temperate climate such as that of Britain, there is also another striking trend which shows a seasonal effect on mortality rates with excess deaths occurring during the winter. This can be expressed as a 'winter mortality ratio' by comparing mortality rates for the January–March quarter with the average for the whole year. Improvement in public health measures in Britain during the early part of the present century effectively reduced the secondary peak of mortality in the summer (due to enteritis, cholera, etc.) and this had the effect of increasing the winter mortality ratio. In the last two decades, however, the ratio has been decreasing, particularly in the older age group. The recent fall in winter seasonal mortality cannot be explained by fewer severe winters or less frequent serious influenza epidemics. The explanation may reside in two environmental changes which have taken place since about 1960. The first is the increased use of central heating. In countries such as Japan and in the U.S.A. central heating is claimed to have a major impact in bringing about a decline in seasonal mortality. This could also apply to the U.K. where central heating and home insulation is grad-ually improving. In the most recent extensive survey of temperatures in domestic dwellings in the U.K. during February–March 1978, centrally heated houses ran 3 °C warmer on average than non-centrally heated houses.

The second environmental factor is the reduction in air pollution due to smoke and other air pollutants. While both of these factors clearly may have an effect in reducing winter mortality it is not easy to prove a causal link. Other considerations may be involved which caution against drawing hard and fast conclusions from statistics available at this stage.

ENVIRONMENTAL TEMPERATURE AND DEATH-RATES

It would be reasonable to suppose that seasonal fluctuations in mortality are determined by seasonal temperature cycles as well as by the types of diseases prevalent in a particular country. Other important considerations which could have a seasonal impact on mortality, especially in developing countries, are the various socio-economic pressures such as impoverishment, diet, unemployment, etc. In more 'advanced' countries, many summer diseases have been more or less arrested, with the peak of disease occurring in the winter months. But even in such countries the winter peak, if related to environmental temperature, could be reduced, for coldness in winter is not an unsurmountable natural obstacle. In the northern states of the U.S.A., Canada, and in Scandinavian countries where winter temperatures drop well below zero, large-scale area heating and efficient housing insulation is the norm. The evidence suggests that an artificially maintained warm climate may contribute to a reduction in winter deaths and a damping down of the seasonal variation in these countries. In the U.K. and Japan, where the winter climate is more temperate and variable rather than consistently cold, housing stock often lacks central heating and room temperatures can be as low as 10 °C or less in winter. The winter mortality ratio is higher in such countries though, as we have seen, the seasonal fluctuation in mortality is becoming less pronounced.

It is interesting to see how Great Britain compares with other temperate countries in terms of seasonal variations in mortality. In Table 2 the number of deaths occurring in a particular month is expressed as a percentage of the number expected if mortality was evenly spread over the whole year. Thus the *average* number of deaths per month calculated from

Table 3. *Monthly variations in mortality from all causes, for temperate countries during the period 1968–72. (From United Nations Demographic Year Book, 1974)*

	Jan	Feb	Mar	Apr	May	Jun	Jul	Aug	Sep	Oct	Nov	Dec	Coefficient of variation
England and Wales	130	114	111	100	93	89	86	84	86	91	99	119	0.151
Scotland	131	113	110	100	94	91	88	85	89	90	97	111	0.135
Denmark	114	106	103	99	98	97	93	92	95	97	101	106	0.064
Finland	110	107	100	96	99	101	97	92	94	97	99	110	0.058
France	114	112	110	101	94	91	92	86	89	94	99	117	0.105
Germany (W.)	117	112	106	99	92	96	92	90	91	95	99	113	0.093
Holland	118	106	102	97	94	95	95	91	93	97	101	109	0.079
Norway	113	103	103	97	98	99	97	94	94	97	100	107	0.057
Sweden	111	104	99	99	95	99	98	93	95	98	99	110	0.055
Canada	114	104	99	97	98	97	99	94	95	98	99	105	0.056
U.S.A.	114	106	100	97	97	96	95	94	93	97	101	108	0.062
New Zealand	90	88	88	94	101	111	120	114	106	99	90	71	0.141

the yearly total is taken to be 100 and the *actual* number in each month is a percentage of this. A high coefficient of variation (Table 3) means that there is a large seasonal effect.

The Table shows higher winter death-rates in Great Britain, particularly in the month of January, compared with other temperate countries in the northern hemisphere. It also shows lower death-rates during the summer so that the seasonal fluctuation appears to be even more accentuated in Great Britain.

An analysis of daily deaths in England and Wales and in New York which was undertaken by Sir Graham Bull a few years ago showed that deaths from heart attacks, strokes, and pneumonia increased linearly as environmental temperature fell from 20 °C to –10 °C. But there was also a steeper rise in mortality rates as ambient temperatures rose above 20 °C and as the temperature fell further below –10 °C. The mortality-ambient temperature relationships were much more marked in elderly people. Another finding of great interest was that short-term temperature changes over one or two days had only a small effect on death-rates but significant effects were noticed when temperature changes lasted for one to three weeks. It was observed further that deaths from heart attacks tended to occur one to two days after, from strokes three to four days, and from pneumonia about one week after the change in temperature.

The numbers of fatal cases resulting from cardiovascular and respiratory diseases during cold winter months in Britain are far greater than those in which hypothermia has been registered as an associated cause of death. It is not appropriate therefore to consider hypothermia as such as the major cause of death in cold conditions. Clear-cut cases of environmentally induced hypothermic deaths are, as we have seen, relatively rare. The statistics which have been shown to be significant are those associated with increased mortality due to cardiovascular and respiratory diseases in cold conditions. The extent to which secondary hypothermia can be considered to be involved in these deaths is clinically almost impossible to judge at present.

5

Hypothermia in the new-born

For the newborn infant, cold is one of the first stresses which greets its entry into the world. This is largely true of all new-born mammals when the young emerge from the protection of the maternal environment and are precipitated into colder and more hostile surroundings. The obstetrician and paediatrician are very conscious of the need to minimize the impact of this sudden environmental change. At birth, homeothermy, the ability to keep one's own internal temperature constant, is put to an immediate test, but whereas the adult can regulate body temperature at a constant level with apparent ease the newborn baby is less capable of maintaining its own temperature, at least for a time. Low birthweight babies are known to be particularly vulnerable to cold and special measures are taken to shield them. Swaddling the newborn infant is of course the human way of protecting the baby from cold at birth just as other mammals protect their young with their own fur or in a pouch or nest. The risk of hypothermia in the human neonate can be minimized by careful nursing and if necessary by the use of specially designed heated cots or temperature-controlled incubators, which can also serve to provide extra oxygen if required. Primitive incubators have in fact been in use for almost 100 years.

THERMOREGULATION IN THE NEW-BORN

The body temperature of the human fetus 'in utero' is higher than that of the mother by about 0.5 °C. Heat is produced in the fetus by its own metabolism but there is no significant channel of heat loss through the fetal skin or from the lungs as there is after birth. The mother's placenta is the main channel for removing the heat produced by metabolism in the fetus.

A pregnant woman who develops a high fever will respond in the normal physiological way by diverting more blood from the core of the body to the skin, and the hot skin is thus able to dissipate heat more rapidly. This relative shift of blood from the core to the skin, however, will occur partly at the expense of blood-flow to the uterus and placenta and it has the effect of reducing the efficiency with which heat is removed from the fetus via the placenta. The result is that the fetus also develops a fever. Similarly if hypothermia occurs in the mother the temperature of the fetus will also be decreased.

After birth, the self-maintenance of a central body temperature within a fairly narrow range between 35.5 °C and 37.5 °C becomes critical for the neonate. A constant body temperature improves survival in the first days of life and when body temperature is maintained in the normal range the rate of growth and development is optimum.

Basal heat production

During the first 10 days or so of life the newborn baby shows a rise in its resting (basal) heat production, and throughout childhood heat production generally remains higher than that of an adult compared on a weight-for-weight basis. On a unit body-surface-area basis, however, basal heat production for the first two or three months is less than that of the adult. This is the result of the physical fact that a baby has a relatively larger surface area per unit of body weight than the adult. Preterm babies have a still larger relative surface area and an even lower basal heat production per unit surface area. In coping with the thermal environment the neonate suffers two major disadvantages due to its size: a lower basal heat production for a given area of surface, and a relatively large surface area for promoting heat loss.

Hypothermia in the new-born

Heat loss from the skin

At birth, even in very small preterm immature babies, thermo-regulatory control exerted by nerves over the skin blood-flow is well developed. This means that the baby can constrict its skin blood vessels to help prevent heat being lost from the skin surface in cold conditions and dilate the vessels to increase heat loss in the warm. However, as mentioned before, the baby has a relatively large surface area from which it can lose heat to the environment. The baby also has a smaller thickness of fat under the skin surface for insulation, less than half that of an adult man, with especially little fat insulation in babies of low birthweight. The baby also possesses a thinner horny surface layer of skin (epidermis) and an increased skin content of water. Thus when the baby becomes exposed to a drier, colder environment in the delivery room, the inherent greater water content of the skin together with the wet surface of the baby allow a large amount of heat to be lost from the skin surface by evaporation. All of these factors promote heat loss from the newborn and if there is not adequate protection by wrapping the baby in blankets and keeping it in a warm room hypothermia may quickly develop.

Shivering

Non-synchronous muscle contractions on exposure to cold, which we recognize as shivering, is an important defence in low-temperature conditions since it increases the rate of internal heat production and raises core temperature. Most healthy babies can double their resting heat production in the cold, but in babies of low birthweight the shivering response is always muted. Even in the normal neonate there is debate as to how much of a role shivering actually plays. Newborn babies rarely appear to shiver, but undoubtedly some muscle activity is responsible for the rise in heat production even if this is not observable as gross movements of the extremities characteristic of shivering. Much of the extra heat produced internally in the new-born when there is cold-exposure is now thought to be due to the metabolic activity of a special tissue known as brown adipose tissue.

Non-shivering thermogenesis

Brown adipose (fat) tissue has been shown to be responsible in a whole range of neonatal mammals, including the human, for non-shivering thermogenesis, i.e. heat production during cold exposure which does not involve the contraction of muscles by shivering. In certain small mammals this special fat tissue is involved in arousal after hibernation and also in adaptation to continuous exposure to cold in the winter. It has been shown that brown adipose tissue is an important source of extra heat production in the human infant during exposure to cold. The tissue is located just beneath the skin around the neck, in the axillae and the upper part of the back, round deeper structures such as the spine, the large blood vessels of the heart and the kidneys. It differs from normal yellow or white fat in its capacity to generate heat when the body is cooled. When an infant is cooled, brown adipose tissue is stimulated to release and burn the fat in the cells and as this fat is combusted, heat is transferred throughout the whole body through the bloodstream.

Although brown fat plays an important part in thermoregulation in the newborn human infant it appears to have little importance in later life. The tissue has been identified in adults but it is often difficult to find and it probably contributes little to total heat production in the cold. As a matter of fact normal brown adipose tissue is not brown to the naked eye and it is difficult to distinguish it from ordinary white or yellow adipose tissue. The brownness may become apparent after fat depletion when there is a high concentration of substrate substances (e.g. cytochrome). The difference between brown and white fat can only really be appreciated under the microscope where it can be seen that brown adipose

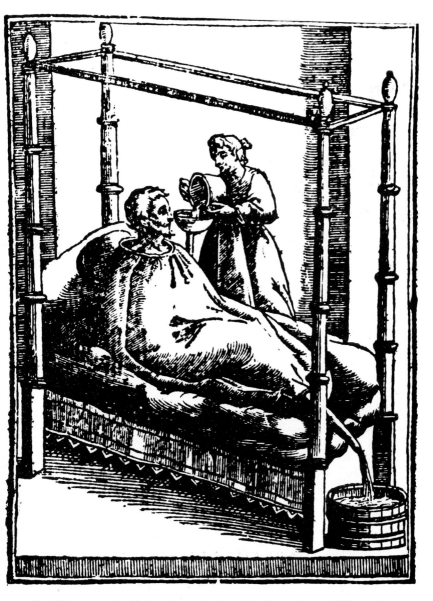

PLATE 1   A cooling blanket designed and used by Sanctorius (*c.* 1660) for the
treatment of patients with fever.

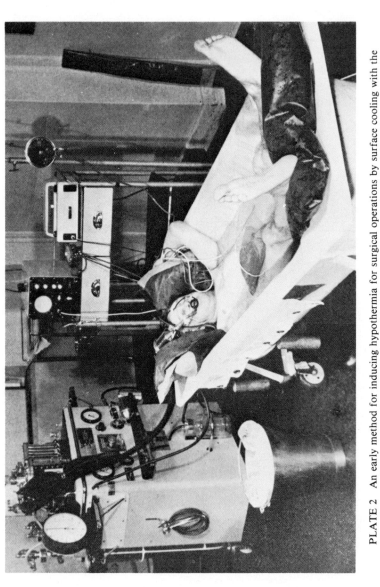

PLATE 2 An early method for inducing hypothermia for surgical operations by surface cooling with the patient immersed in ice-water. It was said that, with correct management, the risk of frostbite was negligible. (Reproduced by courtesy of Mr. Charles Drew, published in *British Medical Bulletin* **17**, 32 (1961).)

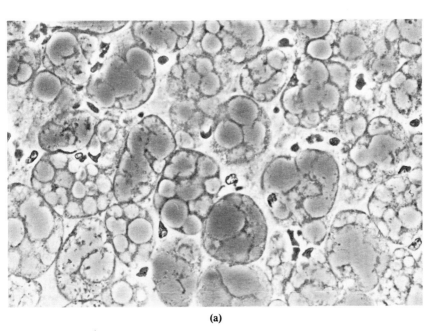

(a)

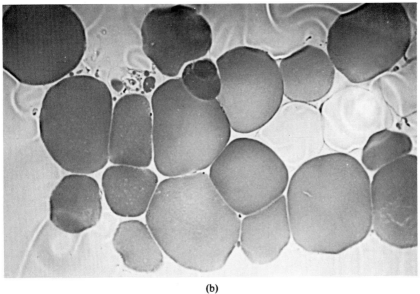

(b)

PLATE 3 (a) Brown-fat tissue from a neonate showing multilocular fat cells with dense cell contents and numerous blood vessels in the tissue; (b) white fat cells from an adult.

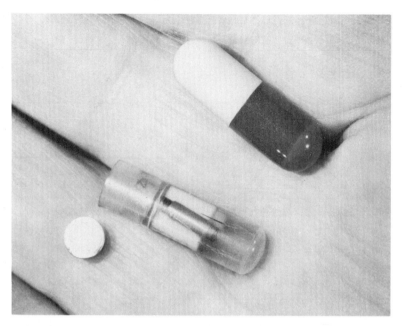

PLATE 4 A radio-pill for recording deep body temperature after the miniature transmitter has been swallowed. The transmitter is powered by a small battery (bottom L) which has been removed for display, and a comparison of size is made with an antibiotic capsule (top R).

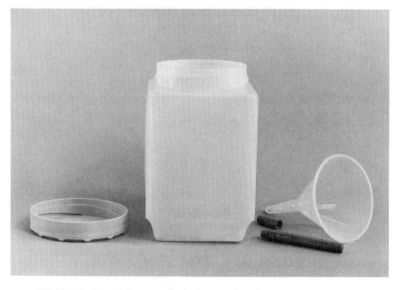

PLATE 5 The Uritemp method of measuring deep body temperature.

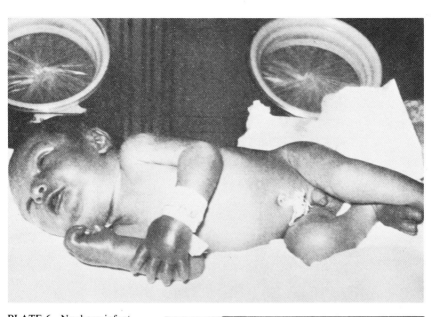

PLATE 6 Newborn infant aged four days with hypothermia. Rectal temperature 28.3°C. Oedema and reddening of the face and limbs is typical of neonatal cold injury. (By courtesy of Dr. C. H. M. Walker, Ninewells Hospital, Dundee.)

PLATE 7 Body cooling unit for inducing shivering responses by cooling in a stream of dry air at 20°C.

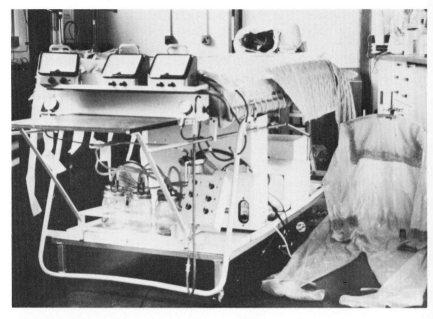

PLATE 8(a) Thermoregulatory function test bed. An air-conditioned garment attached to the bed enables body cooling and warming to take place under controlled conditions.

PLATE 8(b) Two thermoregulatory function test beds in operation.

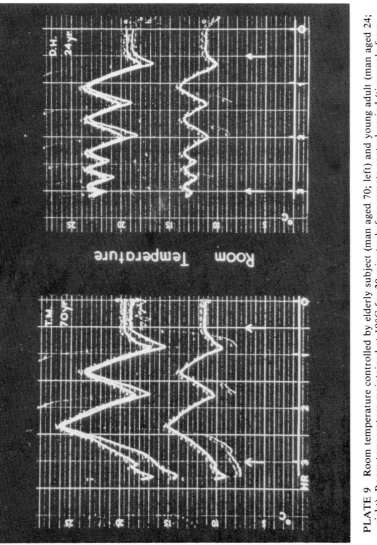

PLATE 9 Room temperature controlled by elderly subject (man aged 70; left) and young adult (man aged 24; right). Room temperature maintained at 19°C for 30 minutes before remote-control period (time scale from right to left). Air temperature measured at two metres above the floor (a), and at table height (b); wet-bulb temperature (c).

PLATE 10 The genesis of hypothermia in elderly housebound individuals. Poverty, poor heating, and a cold, damp environment are often indicators of the elderly at risk. (By permission of Camera Talks Ltd.)

tissue cells contain many smaller 'locules' and have an abundant blood- and nerve-supply whereas white fat is unilocular and contains less dense cell contents (Plate 3). The distinction is often not as clear-cut as this, however, and special studies to locate 'enzymes' in the tissue or high-resolution electron microscopy is sometimes used. The interesting observation has been made that fat in the cells of brown adipose tissue may be depleted in babies who are hypothermic.

Behavioural thermoregulation

With all the disadvantages that the newborn baby, especially the preterm infant, has in controlling its body temperature, it is no wonder that babies are considered to be almost three times as susceptible to fluctuations in environmental temperature as the adult. Perhaps even more of a serious handicap, however, is the baby's inability to alter, to move away from, or even to indicate clearly the presence of an adversely cold environment. For the adult this is where behavioural thermoregulation comes into play, but the baby is not able to control room temperature nor to dress itself in warmer clothes to increase insulation in the cold. It is the adult's responsibility to ensure that the infant is not needlessly deprived of adequate clothing insulation or exposed to a chilling environment.

Unfortunately, the well-clad adult is not always the best judge of the extent to which a baby is suffering from cold-stress. No better illustration of this appears in a recent report from North America where the parents of a one-year old took their infant, well wrapped-up in a sling on the back of their ski clothes, as they skiied cross-country in an air temperature of −6 °C. The parents kept very warm with the activity, even to the extent of sweating. The infant, quite static on its father's back, unable to gain heat from its parent's body through the insulative ski-clothes, was found to be hypothermic at the end of the journey and did not recover.

Hypothermia: the facts

PREVALENCE OF HYPOTHERMIA IN INFANTS

Two of the most serious 'epidemics' of infant hypothermia which have been recorded in the U.K. occurred during the winters of 1961–2 and 1962–3 in Glasgow. The winter of 1962–3 was one of the coldest in living memory. During the two winters at least 110 babies were admitted to Glasgow hospitals with deep-body (rectal) temperature of 32.2 °C or below, the borderline of severe hypothermia. The incidence of hypothermia was not restricted to the colder northern regions of the country. Even before that, in the 1950s, much experience of hypothermia had been gained by paediatricians working in the Oxford area where hypothermic infants were also admitted to hospital in alarming numbers from their homes in the old, cold stone cottages and farmhouses of the Cotswolds.

Hypothermia can occur in young infants as the result of severe chilling caused by bathing in an unheated room, prolonged exposure during drying, and then wrapping the baby up, still cold, to be returned to a cold bedroom. Or worse still, to be put out in a pram in the 'fresh air' of a cold winter's day. Because preterm babies have not developed sufficiently to produce enough heat to combat their heat losses they simply remain cold when wrapped up after becoming chilled, rather as ice-blocks used to be stored in hessian or straw.

Hypothermia is associated in infants with the 'neonatal cold-injury syndrome' which describes the typical appearance of many severely hypothermic infants (Plate 6). The syndrome very rarely occurs in hospital because the environment is usually adequately heated. Most cases are admitted from their own homes. Fortunately, the threat of hypothermia in infants has now been clearly recognized as a winter hazard in cold and temperate countries where there is a poor standard of home heating and a concerted effort has been made to inform parents of the dangers of cold environments.

Hypothermia in the new-born

RECOGNIZING HYPOTHERMIA IN THE INFANT

From inspection, hypothermia is unlikely to be recognized at first because the onset of symptoms is gradual and the baby does not appear to be ill. In fact a most misleading sign often seen in an hypothermic baby is redness of the cheeks and extremities which gives the impression at first glance that the child is perfectly healthy. Rosy cheeks in the cold are the result of blood in the surface skin capillaries remaining oxygenated, and therefore bright red. At low temperatures, blood haemoglobin tends to hold on to its oxygen and the oxyhaemoglobin so formed does not dissociate so readily. This does not mean that every red-cheeked, quiescent baby needs to be rushed to hospital! Diagnosis of hypothermia includes the important observation that the baby's body is very cold to the touch and that there is usually some obvious environmental feature causing the body temperature to drop to a very low value. It is at a later stage when deep-body temperature has reached hypothermic levels that more obvious signs appear indicating that the child is unwell. The baby becomes noticeably lethargic and difficult to feed, the heart-rate and respiration-rate become slow and the whole of the body surface feels cold. In severe hypothermia the limbs and face become swollen (Plate 6) and the tissues just beneath the skin are stiff and hard because of the excessive accumulation of fluid. Infants who have developed hypothermia need to be transferred to hospital for special care, and body temperature raised carefully to a normal level again.

6

Accidental hypothermia

Young, healthy adults can suffer hypothermia as the result of prolonged exposure to cold weather or accidental immersion in cold water. More often than not, an unforeseen event overtakes the victim during outdoor activities in potentially dangerous cold surroundings. Many cases of hypothermia occur on trips that would be considered easy in fine weather but then the weather deteriorates and the situation rapidly changes to one involving a fight for survival. For those not exhausted and who are determined to survive the outlook is often hopeful, but there may be physical injury and immobilization after a fall — a typical setting for accidental hypothermia. As will be seen in a later chapter, accidental hypothermia can also occur in the domestic environment when an elderly person, unsteady after getting out of bed at night in a cold bedroom, trips and falls, perhaps fracturing a limb. The old person may not be found for some time and in the meantime has lain on the floor of the bedroom with little clothing for protection, suffering eventually from exposure and progressive hypothermia.

At sea, shipwrecks, ditching aircraft, and offshore accidents are the common cause of accidental immersion hypothermia, but many other accidents also occur in boating and swimming incidents in inland rivers and lakes. There is much evidence to suggest that hypothermia may rank with drowning as a cause of death following shipwreck in cold water. Drowning is often the consequence of the hypothermic victim losing consciousness. The problem of accidental hypothermia among cavers is also one which commonly involves immersion in cold water, often caused by rain which

Accidental hypothermia

leads to sudden filling of underground streams.

In regions nearer to the poles the problems of cold-exposure are much greater, but appropriate measures usually exist to cope with the cold and there is less of an element of surprise when the weather deteriorates. In the State of Alaska, accidental hypothermia is a year-round hazard and in the town of Anchorage, a 'Department of High Latitude' has been established in order to study all medical problems of the far north, including hypothermia. Clearly the opportunities for research into hypothermia here could be limitless. A specialist on the treatment of frost-bite, Dr. William Mills, has reviewed some 66 cases of hypothermia which have occurred in recent years in a region of Alaska, and examined the causes (Table 4). Twenty of these patients had a deep-body temperature between 19 °C and 27 °C, of which eight died, and 46 patients had temperatures between 28 °C and 35 °C, all of whom survived. Most of the cases of hypothermia listed in Table 4 can be regarded as accidental. The cases appear in all walks of life and in a wide range of human activities.

Table 4. *Causes of hypothermia: a review of cases in Alaska (Data supplied by courtesy of Dr W. J. Mills)*

Cause	No. of cases
Alcoholic stupour	15
Psychotic episode	3
Drug abuse	4
Children abandoned in snow	2
Victim of assault	5
Attempted suicide	2
Vehicle accident	3
Aircraft accident	7
Found in home or cabin	2
Shipwreck	11
Freshwater immersion	4
Lost in blizzard or storm	3
Lost in wilderness	2
Mountain accident or avalanche	3

HIKERS AND RUNNERS

Hypothermia sometimes overtakes the walker, the cross-country specialist, or the unwary hiker. It is often an insidious, slowly developing condition in these circumstances and the interplay between the various channels of heat loss and gain described in Chapter 2 forms the basis for understanding why it occurs. Outdoors in the cold, if the sun is bright, the body may gain some heat from solar radiation. As might be expected, most instances of hypothermia on land have been reported in overcast conditions, in wet weather, or during the night. In the case of a properly dressed individual, heat loss from the body by convection is probably not limiting but evaporative heat loss is very effective and in some circumstances can be highly dangerous. Loss from evaporation occurs primarily because of wind combined with wetness of the outer surface covering the body and if clothing becomes saturated heat is conducted rapidly through the clothing and lost by evaporation from the surface. Clothes should be adequate enough to protect against wetting, to insulate, and to be wind-proof. In wet, cold, and windy conditions inadequacy in these properties can lead to hypothermia even when outdoor temperature is only moderately cold (e.g. 10 °C). Clothing which has been saturated with water provides little insulation. Impermeability of outer clothing is not, however, always desirable for when metabolic activity is high and the individual is sweating, permeable covering will allow sweat to evaporate and may improve both heat balance and comfort. It is important, however, to have available, some form of water-proof covering in order to safeguard against cold and wet conditions.

Conduction of heat from the feet or hands in contact with cold surfaces is usually comparatively small. The extremities will vasoconstrict in the cold and reduce conductive heat loss. Though often causing great pain and numbness, vasoconstriction in the extremities is an important defence against conductive heat loss, but subjects with painfully cold hands and feet may still tolerate hours of exposure to cold with no reduction in deep-body temperature and no shivering.

Another consideration is the size and build of the individual. The slender or slight person has a smaller thermal body mass with which to produce heat but a greater surface to weight ratio which, on balance, makes for a relatively greater rate of heat loss. Body fat provides some protection from the cold on land but it has a much more beneficial effect in water where fat provides insulation against direct conductive heat loss. However, a fat person on land is likely to reach exhaustion more quickly than a thin person who has less body weight to carry. Physical fitness, especially that acquired in the terrain normally encountered by the walker or mountaineer, is highly beneficial, for it usually ensures that activity can be sustained in adverse conditions for longer periods and without fatigue.

The runner, training or competing in long-distance events during the winter, presents a slightly different problem. A balance must be reached between the excessive metabolic heat produced by muscular activity and the heat loss allowed by clothing. Running generates a great deal of heat and the body's reserves of energy are rapidly used. When body temperature rises thermoregulatory responses cause the temperature of the skin to increase in order to dissipate heat. The gradually rising body temperature will also initiate sweating. Sweating always occurs during long-distance running and if the balance of clothing and activity is correct the sweat will evaporate. But if the clothing is impermeable and prevents evaporation, sweat may actually accumulate on the body surface and in sub-zero temperatures may freeze in the clothing and destroy its inherent insulation. A runner creates a wind effect by his movement and this combined with the prevailing wind helps to cool the body even faster. In addition, we must take into account the effect of increased breathing during exercise which leads to increased amounts of heat and moisture being lost with the expired air. The situation for the runner in cold weather, is therefore of massive heat losses caused by vasodilatation, sweating, and a wind effect on the body surface which is offset by a high rate of internal heat production. Should energy reserves become drained and the runner forced by exhaustion and dehydration to stop, then hypothermia can quickly attack.

Recognizing the signs

Recognizing the candidate for hypothermia is not easy in the early stages. The observer himself may be suffering from fatigue and be less aware of accumulating adverse circumstances.

Deteriorating weather conditions in isolated situations should alert the hiker to the possibilities. The weather may be threatening, cold, wet, and with a chilling wind, or darkness coming on. There is the necessity to hurry and probably a long distance still to cover. Improper clothing may lead to sweating, water loss, and dehydration while extra exertion produces physical and mental fatigue. The incentive of competition will sometimes mask these warnings with the result that a severe state of exhaustion and wetness is reached before a rest is called.

In outdoor sports and activities on land in cold conditions, symptoms of developing hypothermia can be recognized when exhaustion gives way to continuous shivering. Not all people shiver effectively, however, and physical activity tends to suppress shivering. The potential victim often displays poor co-ordination and unusual clumsiness with a slowing of the pace, and dazed, confused behaviour. At this stage a halt may be called, and extra heat can be lost rapidly by conduction if the exhausted individual sits or lies on a cold surface such as rock or snow. If there has been sweating, the wet clothing will continue to dry and chill the skin. Heat loss from the surface of clothes is enhanced if rain has fallen or if there has been accidental wetting in snow or streams and a strong wind is blowing at the unsheltered individual. The rate of heat dissipation under these conditions, coupled with wind chill, can be startling — and lethal.

There should be no delay in starting rewarming procedures; the longer the delay the worse are the chances of preventing the dangerous condition of severe hypothermia. Self-diagnosis is unreliable. Just as the mountaineer often does not recognize that he is suffering from lack of oxygen (anoxia), so the hypothermic individual is unaware of the danger presented by his own low body temperature. Anoxia and hypothermia together present a formidable combination to threaten the mountaineer.

Wind–chill

A 'wind–chill index' has been devised as a means for describing the rate of cooling of a person exposed to a cold wind and to try to predict the cooling effect. Air movement and temperature are thus combined into a single index of the cooling power of a dry, cold environment. The relationship between wind speed and cooling power at different temperatures is not linear. It is expressed in figures given in Table 5.

Table 5. *'Wind–chill' related to air movement and ambient temperature*

Sensation	Wind speed (m/s)	(m.p.h.)	Temperature (°C)	Wind–chill index (watts/m^2)
Intolerably cold: exposed skin freezes in 30 seconds	25	55	–34	2900
Exposed skin freezes	10	22	–23	1720
Bitterly cold	5	11	–14	1400
Cold	3	7	0	700
Cool	2	4	+ 10	400
Comfortable	1	2	+ 20	230

Hypothermia: the facts

The scale of the index is said to relate to the heat loss from a person (in watts per square metre of body surface). The index does in fact correspond quite well to experience in intensely cold, windy field conditions and relates to the degree of discomfort and tolerance. It is theoretically impossible, however, to express the effect of air speed on heat loss without taking into account the amount of clothing worn and since this is not specified wind–chill must be regarded as an empirical index only. The index was devised from experiments in Antarctica based on the rate of cooling of non-insulated cans of water. This, in effect, would correspond to the effect of wind–chill on unclothed parts of the body such as the hands and face.

The Four Inns Walk

This annual event was the setting for several tragic incidents of accidental hypothermia among young men, aged between 17 and 24 years, who were competing over a 45-mile walk in the Peak District of Derbyshire in 1964. The course is located in moorland, the walk being at altitudes from 650–2000 feet (195–600 metres). The record time was 7½ hours but the normal time for completion usually varied from 9½ to 22 hours. The event, as usual, took place in March and details of the course, kit required, and advice about clothing was issued to all competitors. Checkpoints were established at 3–8-mile intervals along the route, a rescue team was on call: the organizers had, in short, taken great care to safeguard the participants.

The weather forecast given to the three-man teams setting off at two-minute intervals from 6 a.m. that morning was that there would be showers with bright intervals. In fact, there was drizzle at the start and then the weather deteriorated all day with heavy rain and strong winds (up to 30 m.p.h.) on the highlands; the rain turned to sleet and snow during the night. Temperatures were between 4 °C and 7 °C during the day and

66

this dropped to 0 °C on the moors at night. The environment for exhausted walkers was cold, wet, and potentially dangerous.

At the checkpoint at Snake Inn, some 16 miles from the start, word was received at about 1 p.m. that some competitors were in difficulties on the most taxing part of the route — a long ascent of 1300 feet over boggy ground. The local mountain rescue team went out and brought back five exhausted walkers, two in a state of collapse. One of these, aged 19, was reached at 2 p.m. and was able to walk with assistance but he later collapsed and was eventually transported on a stretcher. According to his team-mates he had begun to flag around mid-day after a tiring climb, 5½ hours after setting out. He began to fall down frequently and had to be supported. When the rescuers arrived two hours later he was conscious and walking, but he soon became semi-conscious and incoherent. By the time he had arrived by stretcher (on which he was protected in a thick kapok sleeping bag with a waterproof cover) at a checkpoint at 7 p.m. his body was rigid and he had become very pale. An hour later he was admitted to hospital and found to be dead. At an inquest following these events it was stated that this young man had suffered a severe attack of influenza three weeks before the walk, a factor which might also have contributed to the fatal outcome.

Another three-man team had started at 7.45 a.m. and by the afternoon found the going more and more difficult. One member complained of cramp and kept stopping. All three had become very wet and had lost their way, being about a mile off course. The member of the group who kept stopping was clearly becoming very tired and unsteady. He had to be urged on by his companions and when he could go no further they sat down and the fittest member of the team went to reconnoitre. After he had returned, the other mobile member of the group went to fetch help and eventually found a rescue party who took him to the checkpoint where he was found to be suffering from exposure and exhaustion and unable to give an exact location of his two lost companions. Two parties of rescuers then set out to find the two men, but failed. It is thought that one of them had stayed with his companion until he died and then had started back in the dark. Their bodies were found two days later, one partly in water in a stream, the other lying covered in snow about a mile away from his companion. The clothing of the young men who had died of hypothermia on the Four Inns Walk was later examined to determine the insulation (clo) value. It consisted of an anorak (two of the three were of poor quality), jersey, shirt, singlet and pants, trousers (one jeans, one 33 per cent terylene–wool mixture, one corduroy — none of which were waterproof) two or three pairs of socks each and climbing boots of good quality. This is a fairly standard assembly, giving an insulation value of 1.5 clo and which, when dry, is enough to keep someone warm when walking at 3 m.p.h. on the level. If the clothing was thoroughly wet, the clo value would drop to well below half the dry value and a heat production in the region of 580 watts, e.g. walking up a 5 per cent gradient at 3–4 m.p.h. would be required to maintain heat balance. However good an anorak may be, it may not keep a person dry for more than two or three hours, for anoraks are often made of semi-permeable cloth to avoid condensation of sweat. All competitors giving evidence at the inquest agreed that despite wearing anoraks they were wet through and very cold. A patrol warden of the Peaks National Park said that he always wore an oilskin on the moors in wet weather.

There are some obvious lessons to be learnt from these and similar incidents of accidental hypothermia. The important guidelines for those who climb or walk in hill country in Great Britain or countries with similar climates are:

1. Dress properly and carry a waterproof coat (a 'plastic mac' might be life-saving).
2. Keep in a state of physical fitness so that you can cope

with prolonged exercise without becoming exhausted.

3. If the weather deteriorates and the destination is not near, 'go to ground', i.e. seek shelter immediately before fatigue sets in.

Management in the field

There is no standard procedure for every case of cold-exposure. Management will depend largely on the prevailing conditions and available help, and when an unconscious hypo-thermic victim is found special care in hospital is urgently needed and medical assistance required at the scene. Much can be done, however, to try to prevent severe hypothermia:

1. In those showing signs of hypothermia it is first of all necessary to avoid further exhaustion and heat loss. Rest must be insisted upon and when the patient sits or lies it should be on a relatively non-conducting surface such as branches or leaves, clothing, or packs rather than on rock, snow, or ice. Some form of protection from the wind and rain is another priority but it may not always be available in the form of natural shelter and it may be necessary to construct a 'lean-to'.

2. If clothing has been wetted by sweat or externally by rain, an impermeable cover will help to prevent evaporation. Additional clothing, even if wet, will help reduce convective heat losses. But when there is no possibility of drying out in a sheltered area the important rule is to prevent surface evaporation from wet clothes by additional cover and to disregard the discomfort of damp clothes. Covering the wet clothes with an impermeable layer might retain say one litre of water in the clothes and to heat this water from 4 °C up to 34 °C (warm skin temperature) in one hour would require 35 watts of body heat. In contrast, if one litre of water was allowed to evaporate from the surface in one hour, 675 watts of body heat would be lost.

Hypothermia: the facts

Ideally, if shelter is available, it is better to remove wet clothing and reclothe the victim in warm, dry garments made up of several layers. Hypothermic patients should be handled gently and it may be necessary to cut the wet clothes off in order to avoid manhandling when the garments are replaced.

A 'casualty bag' is usually carried by rescue teams for protecting and insulating the hypothermic victim during a recovery operation. Above all, the casualty bag must be waterproof and windproof and provide a good thermal insulation, preferably by a double layer of fabric enclosing a dead-air layer. One variety of casualty bag incorporates an inner layer of metallized plastic sheeting which is intended to reflect radiant heat back to the individual placed inside the bag. Radiant heat is an important avenue of heat exchange in space and a principal consideration in the design of space suits. Radiant heat can be transmitted in a vacuum. However, in cold air, heat lost by radiation from the skin surface is less important than convective heat exchange and the metallized layer does not therefore add greatly to the insulating properties of the bag. In fact, when there is condensation from moisture on the inner surface of a metallized survival bag the reflecting properties of the metal lining may be lost.

3. A significant amount of heat can be lost from the head and respiratory passages, especially in cold dry conditions. By covering the head, mouth, and nose with a scarf, heat loss can be effectively reduced and a crude heat exchanger and humidifier thus constructed over the mouth and nose. Cold dry air removes moisture from the respiratory passages and another benefit of this procedure is therefore to help to reduce dehydration.

4. If the individual is not severely hypothermic and not completely exhausted, in the absence of other means of rewarming some movement should be encouraged. The size of the shelter may prevent vigorous movement but body

movements provide a way of stimulating internal heat production, and reversing the trend of falling body temperature. With the patient wrapped in dry clothes and a sleeping bag or blankets to prevent heat loss, shivering and some degree of body activity will gradually contribute to restoring body temperature.

5. Another method of warming is by body contact and this may prove beneficial, particularly if several bodies participate to increase thermal insulation round an hypothermic individual.

6. Surface rewarming, for example, by an outdoor fire or warm compresses applied to the body surface is sometimes considered to be undesirable and counter-productive. The reason for this is that surface warming may produce an 'after-drop' in body temperature (see page 79). There is need for caution in this respect in the severely hypothermic unconscious patient where under field conditions rapid dilatation of surface blood vessels, especially in the limbs, may lead to dangerous effects on the action of the heart. However, in the mildly hypothermic state it may be better to risk a slight temperature after-drop if in the long run the fall in deep-body temperature can be reversed by surface warming.

7. Hot sugared liquids can be fed slowly to the conscious hypothermic victim and he or she should be kept awake and talking if possible. No attempt should be made to give fluids by mouth if he is unconscious.

8. Avoid giving alcohol. Generally, a small amount of alcohol does little harm and may raise morale, but if the hypothermic victim is exhausted, fatigued, and without food, blood sugar levels will be low and alcohol can itself cause a dangerous fall in blood sugar level and lead to further rapid body cooling. The familiar concept of the St. Bernard dog being used for locating and rescuing travellers lost in the Alps and carrying its keg of brandy to revive the victim apparently contradicts this theory. Dogs were

certainly used for this purpose but the kegs carried by the dogs actually contained a sugar solution lightly laced with brandy.

9. Finally, it is best not to leave the conscious hypothermic patient alone unless it is unavoidable. Encouragement is an important component of assisting the victim to recover. There are many accounts of bizarre and irrational behaviour by hypothermic individuals when not supervised, which may rapidly undo all the good accomplished by an attempted rescue operation.

CLIMBERS AND MOUNTAINEERS

Mountaineers sometimes suffer from exhaustion hypothermia due to exposure to cold air at altitude, especially when they are in a physically tired, and hungry state with the body's usable energy stores depleted. The onset of hypothermia and exhaustion in this case is much slower than in immersion hypothermia and often involves gradual changes in mood and personality. At sea-level the effort involved in running for a period of several hours may rapidly deplete energy reserves, but mountain climbing is often sustained over days or even weeks and there is continual stress from physical exertion and exposure to an hostile environment. An analysis of mountain-rescue incidents in Scotland for the decade 1967–77 shows that 14 per cent of 1083 surviving casualties suffered from hypothermia. Of 158 fatal casualties during the same period, 10 per cent of deaths were attributable to hypothermia and 75 per cent to physical injury. Many of those who died as the result of physical injury were no doubt also suffering from the effects of chronic exposure to cold which would have contributed significantly to the fatal outcome.

The high-altitude environment is notoriously unpredictable with rapid falls in temperature, squalls and snowfall, sudden gusts of wind, and loss of visibility which can bring climbing to a halt. Above all there is unending exposure to the climate with

Accidental hypothermia

little natural shelter and the ever-present effects of oxygen lack at high altitude. Hypothermia may develop insidiously with exhaustion, and in the mountains the problems of access and rescue become even greater.

A study of major mountain ascents some years ago found that there were three main reasons why climbers failed to achieve their objectives: that the climbers did not get enough sleep, that meals were unpalatable, and that there was dehydration. All of these factors make for chronic fatigue and once this has occurred the prevention of hypothermia becomes much more difficult. Loss of appetite and dehydration combine to affect the availability of energy for body functions and this can profoundly affect the ability to withstand the cold. Heat loss from the respiratory tract is increased at high altitude because of increased ventilation induced by oxygen lack. With a decrease in oxygen available the efficiency of all body processes diminishes and this includes mental changes such as loss of concentration. Behavioural changes of this sort are extremely dangerous in a situation which demands clear thought and decision. Irrational behaviour in the casual visitor to the mountain who becomes overwhelmed by cold sometimes takes the form of paradoxical undressing ('mountain disrobing syndrome'). The hypothermic victim reaches a point where he will undress and roll over in the snow until consciousness is lost. The following report from the Medical Officer of the Braemar Mountain Rescue Team describes a typical case in the Lochnagar mountain region of the Cairngorms in Scotland.

On 9th April 1978 two young men in their 20's set out to climb the Douglas Gibson gully on Lochnagar. This is a difficult, grade 5 climb but the men were experienced and well-equipped. They were late in starting and were within 10 feet of the summit when, at 6.30 p.m., an avalanche occurred. From here they slid, and fell approximately 600 feet, both being conscious when they came to a halt. One of them suffered only minor injuries, but the other had more severe facial and leg injuries and could not walk. There were no other climbers in the corrie at this late hour and so the mobile man attempted to protect his companion by wrapping him in blankets obtained from the first aid box kept in the corrie, and then he set out to bring help. He alerted the rescue team at 8 p.m., but heavy snow, a pitch dark night and the difficult terrain combined to slow the progress of the rescuers. When they arrived at the location at 1.30 a.m., 7 hours after the accident, there was no sign of the casualty. On looking around, a trail was discovered and followed for about 600 feet down the slope, with discarded clothing scattered along the route. His body was frozen solid when he was found and there was no possibility of resuscitating him.

New visitors to mountains often experience 'mountain sickness', which is a term used to describe the effects of the mountain climate on those who are unacclimatized. Some degree of acute mountain sickness appears to affect every person who ascends to altitudes of 10 000–12 000 feet, especially if the ascent has been rapid. Symptoms vary considerably from person to person but the syndrome includes headache, dizziness, fatigue, shortness of breath, lack of appetite, vomiting, nausea, overbreathing, and insomnia and it comes about because of lack of adaptation to reduced oxygen levels in the air. It clearly poses another threat to the body's ability to withstand hypothermia. Even at a height of 3000 feet (914 metres) most people who are not used to such altitudes will experience shortage of breath because less oxygen is available in the 'thin' air. Mountaineers become acclimatized to this condition when enough time is spent at high altitude and adaptive physiological changes occur which increase the capacity of the respiratory system and increase the ability of the blood to take up oxygen.

IMMERSION IN COLD WATER

Immersion hypothermia is one of the most rapidly induced and dangerous forms of hypothermia, because the thermal capacity of water to extract heat from the human body is very great (25–30 times greater than that of air) and victims may lose consciousness and drown in less than an hour. Shivering occurs extensively and continously at low body temperatures, though it may stop intermittently and lead to dangerous periods of 'basking in the cold'. If victims of cold water accidents had been able to resist hypothermia, drowning might not have occurred. Calculations based on the Registrar-General's figures for England and Wales during the period 1959–63 show that the mortality rate for deep-sea fishermen was more than seven times that for the male adult working population as a whole. Many of the fatal accidents occurring in this group of men result from falling overboard and in the cold fishing grounds of the Northern hemisphere rapid death results from hypothermia.

Hypothermia is a problem encountered also by deep-sea divers, and the North Sea oil-drilling region is a typical setting. The problem is one of balancing heat loss through a diving suit with that of the body's heat production supplemented by heat supplies from the surface. At depths below 200 feet (60 metres) divers must use the technique of saturation diving and breathe a mixture of oxygen and helium gas. Unlike nitrogen, under high pressure helium is not narcotic and can be used at great depths. Unfortunately, helium is a gas which can transfer heat more rapidly than air; at a depth of 1000 feet about 30 times as much heat as air at the surface. Warming the air-supply to the diver helps him to work safely even in cold water but he is dependent on heat as much as he is on the breathing mixture. Heat is usually supplied by flooding the diving suit continuously with warm water. The warm water is pumped from the surface to a diving bell at a temperature around 40 °C and thence to the suit; its temperature can be regulated by the diver. Unfortunately, divers sometimes cool to near hypothermia without any marked sensation of cold and then fail to maintain an adequate temperature in the suit. Some of the unexplained deaths during working dives may have been due to confusion and loss of consciousness from hypothermia when the circulating warm water system was not properly regulated.

Physiological responses

Sudden immersion in cold water causes hyperventilation, an increase in both the frequency and depth of respiration, especially during the first few minutes. This is due primarily to a massive drive to the respiratory control system from cold-receptors in the skin. Most people are familiar with this respiratory response when they plunge into cold water; it invariably invokes a gasp. Gasping is a reflex response to the cold and it enhances the risk of aspirating water and drowning. It also helps to reduce the time of breath-holding under water, which again increases the danger of drowning. During submersion in water below 15 °C maximum breath-holding time is only a matter of 15–25 seconds.

The blood vessels in the skin constrict intensely, especially those in the limbs, but in very cold water (less than 10 °C) the smooth muscle in the blood vessels becomes cold-paralysed and there may be a sudden reddening of the skin as the result of cold-induced vasodilatation (p. 24). In spite of superficial dilatation in this way, the deeper skin vessels appear to maintain their constriction because cooling of the deep body tissues does not accelerate noticeably at the onset of cold-induced vasodilatation.

Submersion of the face in cold water with breath-holding induces the 'diving-reflex', which tends to cause a slowing of the heart-rate and constriction of skin blood vessels. But opposed to this, the heart-rate may also increase with the initial shock and alarm of immersion, then later decrease as the body temperature drops. There is also a loss of fluid from the body due to cold diuresis. Urine flow doubles with immersion in thermoneutral water and increases three or four fold when the water is cold. The cold diuresis is thought to be due to pressure from water on the outside of the body and

76

constriction of vessels in the skin, which increases central blood volume and causes a reflex diuresis. Limb strength and co-ordination deteriorate rapidly with falling temperature of nerves and muscle. In very cold water, swimming becomes difficult to perform, behaviour is irrational, and mental activity disorganized. When combined with fear and initial intense discomfort the plight of the immersed victim is desparate and chances of survival are limited.

Survival

Survival in cold water is variable and many personal and circumstantial factors influence the outcome. A most important physical characteristic improving the chances of survival is the amount of subcutaneous fat, which helps to insulate from the rapid loss of heat to the surrounding water. A larger body size also helps because the greater the body mass the smaller is the relative surface area and slower the rate of cooling. Children cool more rapidly than adults and thin people cool quicker than fat people. Some individuals shiver more readily and others constrict skin blood vessels more intensely.

Channel swimmers provide a good example of the benefit of body fat in water. In 1951, 18 of the 20 swimmers achieved the swim from France to England and were in water of 15.5 °C temperature for periods of 12–20 hours. Records of shipwreck victims show, however, that persons immersed in the sea at 15.5 °C usually live only about five hours and seldom more than six hours. Heat loss in the Channel swimmers is reduced because most are fat or grossly fat and their buoyancy in the water enables them to maintain a steady heat production while swimming, in spite of increased body weight. Once fatigue occurs and heat production diminishes, even Channel swimmers cannot remain in the water. The procedure of spreading grease (lanolin) over the surface of the body marginally improves insulation but several pounds of weight are added in order to produce a layer with an average depth of only 2 millimetres. There may be other advantages of greasing, however, such as reducing friction in the water.

The behaviour of people suddenly immersed in cold water has an important bearing on their survival. Active swimming, apart from short bouts of swimming in order to avoid being drawn into the undertow of a sinking ship or to reach a life-raft, is best avoided since it helps to dissipate heat from the body surface. In cold water the best chance of survival is gained by simple measures to stay afloat in one place, and keeping the head and neck out of the water, and the victim should not move or swim about. Water at 0 °C is almost twice as viscous as that at 25 °C so the work required to make swimming movements is increased and exhaustion occurs more rapidly. Heat loss can also be minimized by adopting a curled-up, arms-folded, legs-together posture, and if there are three or more persons in the water to huddle together.

It is often believed that clothing has a harmful effect in water by hampering movement and increasing weight. To some extent this is true, but clothing, though saturated, still provides some insulation which in very cold water may be life-saving. A large variety of survival garments are now available which provide thermal protection. The most efficient are 'dry suits' which keep the wearer dry beneath the suit and surround the body with closed-cell foam which also improves buoyancy. When combined with thick clothing underneath, dry suits can provide insulation which matches the blubber of marine animals and protects for six hours in water at 0 °C. Whole-body 'wet suits' allow a little water next to the skin, but the water is not rapidly circulating and the body is able to warm this film of water which acts as a barrier. This type of survival suit will protect against hypothermia for about three hours in water at 10 °C. Personal flotation devices (life-jackets) provide extra buoyancy and some types also give nominal thermal protection. Their primary purpose is to assist in keeping the person afloat, but some life-jackets are designed to rotate the unconscious person's body from face down in the water to an upright position. Flotation devices keep the head and neck areas of heat loss out of the water and therefore help to delay the onset of hypothermia.

Temperature after-drop

In all cases of hypothermia the essential objective is to restore body temperature to normal by rewarming. There are many techniques for rewarming, and this is discussed in some detail in Chapter 10, but the manner and rate at which body temperature is brought up to normal is often crucial. In exposed and isolated situations there may be no chance of actively warming the hypothermic casualty and all that can be accomplished is to try to prevent any further drop in body temperature by whatever means are available. Once severe hypothermia has occurred, even rescue with the provision of plenty of insulation and waterproofing does not guarantee survival. The example given of the events in the Four Inns Walk on p. 66 clearly emphasizes this.

One of the immediate difficulties of rewarming procedures is the phenomenon of 'after-drop'. It refers to the fact that once rewarming has started, and this may take place as soon as an immersion victim is removed from the water, there will nearly always occur a sudden drop in deep-body temperature for a period of about 10–15 minutes before it starts to rise. This is a serious matter, for a further drop of 1 °C or 2 °C to occur with core temperature already initially as low as 30 °C, may drop body temperature to the level at which the heart will be markedly slowed and irregularities such as ventricular fibrillation (see p. 38) occur. The best hope for survival under these circumstances is for rewarming to be supervised carefully with full hospital facilities for defibrillation and resuscitation.

The reason for the after-drop in body temperature has, since the early investigations on hypothermia, been attributed to the redistribution of cold blood held in the skin which is shunted into the core of the body when the surface is rewarmed. Another explanation has been put forward in recent years as the result of tests with physical models of the body. These tests suggest that there is a time-lag in the temperature change in deep layers of the body during warming after cold immersion. It appears in fact that both mechanisms correctly explain the after-drop, that they usually act together, but that each may be more dominant in different circumstances. In immersion hypothermia where there is rapid conduction of heat from the core to the skin surface the cooling gradient factor during rewarming is probably dominant. With surface cooling in air the role of the skin circulation becomes more important. We shall return to further consideration of the part played by the after-drop in rewarming when reviewing the methods of treatment of hypothermia.

7

Hypothermia caused by drugs

As long ago as the second century A.D. physicians were aware that certain drugs were capable of reducing the temperature of the body and they used them to treat patients with fevers. It is important to recognize, however, that there is an essential difference between drugs which help to reduce a raised body temperature (fever) to normal (e.g. antipyretics such as aspirin) and those which will lower temperature from normal down to hypothermic levels. In the 1950s, development of cooling techniques for deliberately lowering body temperature for surgical operations created renewed interest in therapeutic hypothermia. It was realized that hypothermia, because of its effect in diminishing metabolism and nervous activity, might be of benefit both to the patient and anaesthetist since it reduced the need to anaesthetize patients to a deep plane of anaesthesia with a powerful anaesthetic. The aim was to secure the benefits of hypothermia during surgical operations by using drugs to induce a state resembling that which some animals acquire naturally by hibernation.

This appeared to be a possibility when a new drug, chlor-promazine, was introduced and shown to possess properties for producing hypothermia. The drug formed the basis of the 'Lytic cocktail' developed in France to induce 'artificial hibernation' simply by means of an intravenous injection. Artificial hibernation is an inexact term for the state of hypothermia caused by chlorpromazine, and in fact there is no real similarity between natural hibernation in animals and the effects of the drug. The lytic cocktail consisted of a mixture of chlorpromazine, a barbiturate, and pethidine (meperidine), and a number of attributes were claimed for it. It caused

81

Hypothermia: the facts

'central nervous dissociation' with altered consciousness, reduced autonomic nervous activity, relief of pain, and a dramatic fall in deep-body temperature. For surgical operations following intravenous administration of the lytic cocktail all that was required for effective general anaesthesia was nitrous oxide gas and a neuromuscular blocking substance to produce muscle relaxation. Pharmacologists found it difficult to study the actions of this mixture of drugs. It was difficult enough to determine the specific effects of chlorpromazine itself, as it had very many different properties, apart from its combined action with the other constituents.

The most frequent way by which drugs cause hypothermia is by depressing the activity of the central nervous system, which may impair consciousness and may also affect the involuntary control of thermoregulation. This dual effect often occurs as the result of deliberate self-poisoning with drugs, and hypothermia can develop even when there is no undue exposure to cold. A potentially dangerous situation exists with central nervous depression in a cold environment. Anaesthetists are fully aware of this problem and the possible circumstances leading to hypothermia. The use of anaesthetics and muscle relaxants together has a tendency to cause the body temperature to fall and operating theatres are kept at a raised temperature because of this.

Sedatives are used to quieten a patient, and hypnotics for inducing sleep. Both act on the central nervous system, but if given in large doses or taken accidentally or deliberately in overdose, depression of thermoregulatory control can lead to hypothermia. Various tranquillizing and antidepressant drugs have also been found to have the potential for lowering body temperature. All of these classes of drugs are commonly prescribed for elderly patients and this must be a matter for concern to the physician attending the aged, particularly in the winter months.

In Chapter 2 the physiology of thermoregulation in the cold was described, and this forms the basis for understanding how certain drugs can cause hypothermia. There are a number of ways by which drugs can interfere with thermoregulation, both indirectly and in the long-term. Direct means by which drugs can affect thermoregulatory processes are:

(1) by interfering with the function of the control centres in the brain (and spinal cord);
(2) by influencing 'higher' nervous system functions controlling behavioural thermoregulation;
(3) by depressing the metabolic activity of body cells and inducing an overall fall in heat production;
(4) by inhibiting muscle contraction and shivering;
(5) by preventing vasoconstriction (see p. 13) in the skin and enhancing heat loss from the body surface.

Some drugs, such as alcohol, have an effect on several or all of these processes together and in the presence of cold stress, hypothermia is accelerated. So hypothermia is an added reason to be pessimistic about the prospects of the drunk who falls overboard into a cold sea.

ALCOHOL

The most important action of alcohol is on the central nervous system, where it causes depression of nervous activity. In this respect it acts like an anaesthetic and with increasing doses all the stages of general anaesthesia can be reached. The often-held view that alcohol stimulates and may cause hyperactivity expresses the fact that inhibitory actions in the brain are depressed and restraints removed.

The anaesthetic effect of alcohol on the brain alters thermo-regulatory control in several ways. The first functions to be lost are the finer grades of judgement, observation, and attention so that complicated decisions involved in behavioural thermoregulation are not pursued. This may take the form of stepping out for a breath of fresh air in very cold wintry conditions without extra clothing. Or perhaps the intoxicated individual decides to cool off by diving into cold water. Many victims of accidental hypothermia due to cold exposure or cold-water immersion have been found in a snow bank or the river with evidence of extremely high levels of alcohol (e.g. blood levels of 200–300 milligrams per 100 millilitres). Alcohol may also affect the finely balanced involuntary centre of thermoregulation in the brain. The important way in which this may alter the physiological response to cold-stress is by reducing shivering. Alcohol can also have a dramatic effect on the subjective sensation to cold. Subjects who were immersed in water at 15 °C for 30 minutes after drinking alcohol reported that they felt much warmer than during a similar immersion experiment without alcohol. Victims of shipwreck are likely to find that immersion in cold water is less unpleasant for them if they have been drinking. But the dangers are greatly enhanced if rescue is not at hand. In long-distance swimming a lack of cold sensation will give a false impression of the ability to continue and exhaustion followed by profound hypothermia can then occur.

Once alcohol has been ingested, some of it is absorbed rapidly from the stomach, because it is highly soluble and diffusible, but most is absorbed later in the small intestine. The rapid movement into the blood causes a prompt vasodilatation, especially in the superficial blood vessels of the skin. It is thought that this response is due to a depressant action of alcohol on the vasomotor centre in the brain-stem which normally maintains the skin vessels in a state of vaso-constrictor tone. Vasodilatation causes flushing of the skin and a feeling of warmth, even though the environment is cold. There is a danger therefore that excessive heat may be lost from the skin. In practice, moderate amounts of alcohol are found to have only a small effect on heat loss in cold water and this may be due to the fact that intense cold stimulation of the skin may override the vasodilator effect of alcohol to cause a net vasoconstriction. In air, with a less intense cold stimulus, alcohol may be more effective in inducing vasodilation.

Hypothermia caused by drugs

When people come into a warm indoor climate after exposure to cold, the consumption of alcohol will give the impression of warmth because of vasodilation, but at the same time body temperature may decrease for a short time because of an 'after-drop' effect (see also p. 79).

Another important action of alcohol is on carbohydrate metabolism in the body. Initially blood glucose may increase due to reduced uptake of glucose by the tissues. This stimulates an increased output of insulin which in turn can produce a state of hypoglycaemia. It has been found that alcohol can reduce the input of carbohydrate available for heat production and this continues until all the alcohol consumed has been metabolized. Severe hypoglycaemia is often a feature of acute alcoholism, especially in fasting subjects. Such a depletion of available energy sources in the body greatly increases the seriousness of the situation in cold conditions, when exhaustion and dehydration may also be present. This might well apply to the hill walker who takes alcohol before setting out in cold weather or to swimmers entering cold water. The risk can be diminished to a large extent by eating a meal with the alcohol so that reserves of carbohydrate remain high during cold exposure. Lack of availability of sugar for energy production resembles the metabolic disorder caused by reduced insulin production in the patient with diabetes mellitus. Not surprisingly, it is well established that diabetic subjects are more prone to hypothermia.

Chronic alcoholics frequently show evidence of severe liver damage which profoundly affects the body's metabolism. Complications of alcoholism are seen in altered neurological function and these basically reflect overall metabolic changes rather than the direct toxic effect of alcohol on the nervous system. One of the complications of alcoholism is to produce a deficiency of vitamin B_1 (thiamin). Sometimes this results in a syndrome known as Wernicke's encephalopathy which usually presents as an acute psychiatric illness, incoordination, and disorders of eye movements. These patients have been found to have a limited ability to vasoconstrict in cold conditions and they are also prone to hypothermia.

PHENOTHIAZINES

This important group of drugs, of which chlorpromazine (Largactil) and promazine (Sparine) are examples, may be described as neuroleptics, that is they have an effect in quietening emotional disturbances and are used therapeutically in psychotic disorders. Chlorpromazine has many different pharmacological actions, the mechanisms of some of which are still obscure. Nevertheless, it is an extremely useful drug in psychiatry and is used particularly in the treatment of schizophrenia. In surgery it has been used in premedication because of its central sedative effect, its ability to suppress shivering, and its vasodilating action. It is precisely because of these actions that it was found useful, in higher doses, to lower body temperature and, as previously mentioned, formed the basis of the lytic cocktail. Since its action is long-lasting, chlorpromazine is better given during premedication rather than as a supplement during anaesthesia. There is evidence that, in high doses, chlorpromazine acts in the hypothalamus as well as in other parts of the brain and in this way can have an effect on central thermoregulatory control.

The phenothiazines have a powerful effect in reducing body temperature in conditions when patients are vulnerable to cold. Thus in hypothyroidism, when stimulation of heat production by the cells of the body by thyroid hormone is low, a single dose of chlorpromazine can produce profound hypothermia. Administration of phenothiazines to elderly people requires great care because of the vulnerability of the aged to hypothermia.

HYPNOTICS

Drugs which induce sleep have been widely prescribed for

many years and they are generally safe except when taken in overdose. Barbiturates were once the first line of drugs used to treat insomnia but are used much less nowadays for this purpose because of drug dependence, which can become a serious problem. Non-barbiturate drugs such as nitrazepam (Mogadon) are much more frequently used as hypnotics because they are safer than other drugs and dependency is less than with the barbiturates.

An overdose of a hypnotic can induce hypothermia very readily. A deadly combination can arise from mixing alcohol and barbiturates in the presence of a cold-stress, since alcohol and barbiturates augment each other's hypothermic capability. Many of the hypothermia casualties which occurred after the *Lakonia* caught fire and sank near Madeira in December 1963 may have been due to the effects of sleeping pills taken by people before they entered the water. It was reported that a number of passengers had apparently taken hypnotics, because there was difficulty in arousing some of them during the night the fire started on board. Again, the elderly should be regarded as being particularly at risk with hypnotics. Hypothermia has been reported after as little as a single 5 milligram dose of nitrazepam given to an elderly person of 86 years, despite the fact that environmental temperature was a warm 27 °C. Such cases, however, even in elderly people are rare.

OTHER DRUGS CAPABLE OF INDUCING HYPOTHERMIA

Hypoglycaemic agents

Diabetic patients who are not totally dependent on insulin are sometimes treated with drugs (e.g. biguanides, sulpho- nylureas) which have the effect of inducing hypothermia. Their action may be due to a combination of effects: central impairment of thermoregulatory control, overbreathing, and sometimes profuse sweating, all of which accelerate body cooling.

Antithyroid drugs

The purpose of these drugs is to control overactive thyroid gland activity until a natural remission takes place or the thyroid tissue is removed by surgery. The condition of low thyroid activity (hypothyroidism) is often associated with low body temperature due to a reduction in metabolic rate, and antithyroid drugs therefore possess the potential for causing hypothermia.

Lithium

This drug is used in the treatment of psychological disorders and among its actions on the functions of the brain it has been reported that the thermoregulatory centres may be affected. In some patients it has been observed that the combination of lithium and diazepam (Valium) can induce a state of hypothermia whereas neither drug alone produces an effect.

Vasodilators

Drugs which profoundly reduce vasoconstriction in the skin blood vessels have the capability for producing hypothermia by allowing excessive heat loss from the body surface. Agents which block the nerve endings in the skin or the nerve junctions controlling involuntary function (ganglion blockers) which were once used for treating high blood-pressure, can cause hypothermia in cold environmental conditions.

Cannabis

Hallucinogens such as cannabis have been reported to cause

hypothermia, though there have been few scientific studies of this effect. It is postulated that cannabis can increase sensitivity to cold so that the subject shivers sooner and fatigues more readily. The effects may also include changes in the excitability of nerve cells involved in thermoregulation.

ADMINISTRATION OF DRUGS DURING HYPOTHERMIA

Therapeutic hypothermia with drugs is a technique little used at present by anaesthetists in Great Britain. More often hypothermia is an unwanted side-effect of certain drugs. It can also follow when large amounts of cold blood or other fluids are infused intravenously, particularly during exchange transfusions. However, it is equally important to consider the effects of drugs given when the body temperature is already low or when the patient is suffering from secondary or accidental hypothermia.

Hypothermia depresses metabolism and recovery from an overdose of drug may be delayed because of this. It has even been suggested that there might be some advantage, for absorption of drugs from the alimentary canal is impaired during hypothermia and this sometimes acts to prevent serious poisoning when an overdose of drug has been taken. Most cases of drug overdose require general supportive measures and will rewarm spontaneously from lowered body temperature. Sometimes, however, the combination of severe overdose and hypothermia is enough to prevent any rise in temperature for some time and active warming procedures may then be necessary. Care is needed in raising body temperature under these conditions because the action of drugs already in the body may be exaggerated or 'unmasked' during the process of warming.

Similarly, great caution should be exercised when drugs are given to a hypothermic casualty, because marked and unexpected complications may arise. For example, it might be considered necessary to administer insulin in some types of emergency. In the hypothermic, the first injection will often have no effect, and a second injection might be given, still with no effect. When the patient is warmed, the double dose of insulin which has not been utilized because of hypothermia may suddenly become active and lead to a dangerous lowering of the blood sugar level.

8

Temperature regulation and hypothermia in the elderly

Now King David was old and stricken in years; and they covered him with clothes, but he gat no heat
Kings I: 1, i.

It has been known since antiquity that elderly people have a low metabolic heat production and are less able to keep warm even when well insulated. Succeeding verses from the above passage in the Old Testament go on to describe how hypothermia might have been managed some 2000 or more years ago. The method, if nothing else, was colourful, but apparently in the case of King David it proved to be ineffective.

Most people in their 60s do not like to think of themselves as old; just exactly who is included in the term 'elderly' is often not defined nor understood. One definition describes the elderly as those of 65 years or more, 65 being the pensionable age for males in many occupations in Great Britain. This definition would also apply to women despite the fact that 60 is the corresponding retirement age and that generally women enjoy a longer lifespan than men. If we use such a definition, then it must be recognized that there are large discrepancies between the chronological and biological age of different people. Some people in their 90s are biologically relatively young for their years, and vice versa. Shakespeare's concept of the seventh age of man — sans teeth, sans eyes, sans taste, sans everything — is not particularly apt in modern society supported by social and medical care, imperfect though this may be. The elderly are vulnerable, but old age is not a disease and old people not a special class of invalid. They are an integral and valued part of society and usually do not relish the idea of complete dependence on others. Nevertheless, older people have their limitations. They are often less sure in their movements and balance, and also may suffer some blunting in the special senses such as sight and hearing. There are other physiological systems which may show age-related changes even in the apparently healthy elderly person. One of these is the thermoregulatory system, and many references have been made in this book to the fact that impaired thermoregulation in the old increases the risk of hypothermia. How clearly we can draw this conclusion from research will now be considered.

THERMAL LABILITY IN OLD AGE

A number of finely tuned nervous and chemical controls ensure that body temperature is kept constant over a wide range of environmental temperatures. Thermal lability, i.e. an inability to maintain temperature homeostasis, occurs for two main reasons. First, thermal control of the body may be unsatisfactory because of reduced capacity of the thermo-regulatory system itself, either through inefficiency or disease, or else because of the presence of drugs which interfere with normal regulation. Secondly, external temperature factors arising from the environment may be too severe for the thermoregulatory system to cope with. A moderately cold or warm environment may be well-tolerated by a young person but it may be too severe for an elderly person if his internal thermoregulatory mechanisms are depressed. Two well-known examples of thermal lability in old people testify to this fact. During the coldest winter months in Great Britain there is usually an increase in the number of cases of hypothermia admitted to hospital, especially in elderly patients. In central and southern states of the U.S.A. it is the summer epidemics of heat stroke in the hottest months which are of concern, and this usually involves elderly city dwellers more than the younger population.

92

BODY TEMPERATURE AND HEAT PRODUCTION

Metabolic rate is lower in older people, whether this is measured as heat produced per kilogram of body weight or per square metre of body surface. Lowered metabolism is a primary contribution to hypothermia in disease (e.g. in hypothyroidism) as is the presence of severe cold-stress. The reason why internal heat production is less in old people is a fundamental one. The proportion of body mass made up of functioning cells is usually slightly smaller in the elderly and this results in an overall decrease in the total heat production of the body. In addition, the more sedentary life-style of the older age group results in a smaller contribution to overall heat production from muscular activity while a smaller energy intake decreases the extra heat produced by the 'specific dynamic action' of food. It is necessary, however, to compare metabolic rates in the basal state i.e. when there is complete physical and mental rest. The Mayo Foundation gives figures for normal standards of basal metabolic rate (Table 6).

Table 6. *Basal metabolic rate in relation to age*

Age (years)	B.M.R.(watts per square metre)	
	Males	Females
2	64	61
10	55	53
20	48	42
30	45	41
40	44	41
50	43	40
60	42	37
70	40	37

Because basal metabolic rate decreases with age it does not necessarily follow, however, that deep-body temperature decreases. The heat balance equation (p. 17) suggests that normal body temperature can be maintained at a steady level if a lower metabolic heat production is compensated by a reduced rate of heat loss. There is, however, a tendency in the old for there to be a relative increase in heat loss from the body surface due to a high surface area–body weight ratio in a frail person and sometimes an inability to constrict the blood vessels in the skin. So not only is there a smaller heat production but a tendency for greater surface heat loss.

There is not much information on differences in deep-body temperature between the young and old in resting, neutral conditions. Some studies claim that there is no evidence for alterations in 'normal' body temperature related to age. The most common method of recording body temperature is by using an oral measurement. But there is sometimes difficulty in recording oral temperature in an old person who is edentulous and where there is a greater likelihood of mouth-breathing. This could result in a false impression, in more than one sense, of lower resting body temperature in the elderly. But there have been other studies using more reliable methods of temperature measurement such as urine, rectal, or 'servo-controlled' ear temperature methods. In over 200 young adults using these procedures the mean deep-body temperature was found to be 36.95 °C (standard deviation 0.24 °C). Under the same conditions, studies on more than 200 elderly people aged from 65 to 90 years showed a mean resting deep-body temperature of 36.73 °C (standard deviation 0.25 °C), more than 0.2 °C lower than in the young.

If this is true, then it is necessary to explain why this difference exists. In the tests, the small difference in body temperature could have been due to the inclusion of some elderly subjects with as yet undiagnosed disease, or there may have been greater physical activity or food intake by the young adults before the test. On the whole, it appeared that adequate time was allowed for all those tested to settle down to a basal level and that other differences due to time of day, environmental temperature, and recording methods were standardized. Since the body temperatures were stable, it suggests that the central control mechanism in the brain may have 'set' the body temperature at a slightly lower level in the elderly. There is, in fact, some precedent for considering a re-setting mechanism from the evidence of research on temperature control and investigations of fever, acclimatization, and other changed states of thermal balance. The mechanism is not, however, fully understood. It might be more economical for the body to operate at a slightly reduced basal temperature in old age.

Another body temperature, skin surface temperature, also appears to be different in old people. Skin temperature depends partly on the amount of warm blood circulating through the skin. In the elderly, the blood-supply to cells and organs is diminished slightly and this is true also for the skin. Mean skin temperature likewise is somewhat lower in the elderly. There may, however, be differences in response of the skin vessels to cold in young and old. There is evidence that some elderly people do not constrict surface vessels very effectively. Under these conditions skin temperature in young people could drop lower than in the elderly and increase the temperature gradient between core and skin temperature. Conversely, with a smaller core–skin gradient, elderly people would have a less effective barrier against heat loss.

SHIVERING

Shivering is an important mechanism which is brought into play in a cold environment to prevent body temperature falling. Different skeletal muscles contract vigorously out of phase with each other with the result that there is a great deal of heat produced but little co-ordinated movement. Most of the evidence we have suggests that the shivering process is absent, reduced, or less efficient in elderly people. But the situation is not quite so one-sided as this for there are considerable individual differences in the ability to shiver by both young and elderly people.

During cold-exposure there are usually three phases of the shivering reaction. At first there is a marked increase in muscle tone (experienced as an increase in muscle tension) and this becomes progressively more intense until shivering commences. Then, as the body becomes colder shivering increases until there is a generalized rigor. Finally, if the exposure is prolonged, the shivering and increased muscle tone may disappear and the individual becomes conscious of a general muscular relaxation and feeling of warmth ('basking in the cold'). At this stage body temperature falls rapidly and mental processes become affected and eventually there is a loss of consciousness. These phases have all been studied in man and many fundamental facts about shivering established. Shivering results in an overall increase in metabolic heat production as work is performed by the muscles, and this can be measured as an increase in overall oxygen consumption by the body. The electrical activity of the muscles themselves can be recorded by electromyography in order to analyse the pattern and development of bursts of shivering in different muscles. Most important of all, it has been found that there are two major 'drives' to shivering in the cold, a low skin temperature and a low core temperature. To some extent each can act independently.

Some early investigations in the United States in 1955 found that older people in the age range 52–76 years, sitting unclothed for up to 30 minutes in a 10 °C environment with minimal air movement, did not shiver or complain of feeling cold; younger subjects in the same conditions all shivered and felt uncomfortable. More detailed investigations of these differences have been made in London using a body cooling unit (Plate 7) in which shivering can be reached quickly by convective cooling in moving dry air at 20 °C. This showed quite clearly that shivering ability was not lost in all people with ageing, even in those over 80 years old. There were, however, some changes in the character of their shivering response. For example, the high peaks of muscle contraction achieved during shivering by young people were usually not attained by the elderly and there was often a longer time required to initiate maximum shivering. Most interesting was the fact that though many elderly people did not shiver during the mild cold-stress, some shivered quite well; and the same pattern occurred in the young; some shivered others did not. Another interesting observation was that both in the young and old, the lack of a shivering response was compensated for by a marked vasoconstriction of the skin blood vessels. We can infer that if, as sometimes occurs, there is a loss of efficiency in the vaso-constrictor response in some elderly people then it is more likely that shivering is resorted to as a defence mechanism.

It appears, therefore, that shivering efficiency is reduced in some, though not all, elderly people, and that this may be due to a damping effect on the speed and intensity of the response. A contributory factor may also be a loss of power in the muscles themselves. Shivering is a primary defence against cold but so also is the vasoconstrictor reaction in the skin, and in moderately cold environments the two mechanisms can give mutual support. If one is damped down the other appears to compensate. When body temperature drops to the hypo-thermic level, however, shivering is likely to occur in both young and elderly.

For the adult, non-shivering thermogenesis does not seem to play a major role in thermoregulation though it is an effective means of heat production in the newborn. If there is any con-tribution to internal heat production from brown fat in later years it is likely to be small for though a few cells of brown fat have been found in tissues of younger adults, they disappear almost entirely by the eighth decade. Despite this, brown fat continues to hold the attention of research workers. One group has recently reported that deposits of brown fat could be found in outdoor workers (lumberjacks, painters, timber-men) in northern Finland but not in indoor workers of the same age. The inference is that people who spend much time out of doors in winter may develop non-shivering thermogenesis as a cold-defence mechanism.

VASOMOTOR RESPONSES

In the zone of vasomotor control, i.e. in the intermediate zone of body temperature between the points at which shivering or sweating occurs, vasoconstriction or vasodilatation of the skin blood vessels forms the first line of resistance to temperature stress. Several researchers have shown that the elderly as a group do not have normal vasoconstriction patterns. Again, this fact needs some qualification. Measurement of blood-flow in the wrist and the hand by venous occlusion plethysmography, a technique which demonstrates vasoconstriction or dilatation in the skin, suggests that the elderly have a number of different patterns of vasomotor response to temperature. This can be investigated by a 'thermoregulatory function test' using an air-conditioned bed (Plates 8(a), (b)) to vary the temperature of the air surrounding the body. The normal response, i.e, with rapid vasoconstriction on cooling and vasodilatation on warming occurred in all the young and most elderly volunteers. The blood vessels of some elderly people (perhaps 20 per cent of the total) do not constrict significantly on cooling. Furthermore, in longitudinal studies on a group of healthy elderly volunteers, the proportion showing non-constrictor responses to cooling were found to increase during the course of a four-year period, and this was even more marked after eight years.

A 'non-constrictor' response does not mean that there is absolutely no constriction of the skin vessels on cooling, but that blood-flow to the skin is reduced less than normal and that it takes longer for constriction to develop. Blood-flow studies have mostly been carried out on the hand, which has a good vasoconstrictor nerve-supply. If lack of vasoconstriction

98

occurs over the whole body surface then one would expect tolerance to cold to be significantly impaired.

In most people it is possible to demonstrate transient bursts of vasoconstrictor activity, occurring a few times each minute. This rhythm is one which is generated by the control system in the brain because electrical recording from nerves supplying the skin blood vessels show a similar pattern of waxing and waning response. With the additional stimulus of cold, the rhythmic activity increases in frequency until the vasoconstrictor response becomes continuous. In some elderly people this rhythmic activity is absent or difficult to detect and it suggests that the sensitivity of the vasomotor system is altered.

Another aspect of blood-flow control can be demonstrated during rewarming after cold exposure. One of the reasons for an after-drop in temperature when rewarming follows a mild cold exposure in air (p. 80) is a redistribution of cold blood from the skin as the blood vessels start to dilate. It might be expected that if blood vessels did not constrict very well in the cold, then on rewarming, the after-drop in body temperature would be reduced. This is found to be the case in elderly people who vasoconstrict poorly in the cold.

TEMPERATURE PERCEPTION

Information about the temperature of the environment is fed from thermoreceptors in the skin to the brain where it is interpreted (1) as a temperature sensation of warmth or cold, (2) to give an impression of comfort, whether pleasant or unpleasant, and (3) to bring about involuntary thermoregulatory responses, such as shivering and vasoconstriction. Each of these three responses is processed in the brain but each can be separated from the other. If a warm stimulus is applied to the hand of a hypothermic person the sensation will be considered pleasant and a cold stimulus will feel unpleasant. When applied to someone who is very warm (hyperthermic) the warm stimulus will, on the contrary, be unpleasant; the cold pleasant. The feeling of warmth or cold is here affecting comfort sensation, giving an agreeable or disagreeable impression. However, in both the hypothermic and hyperthermic person the warm stimulus will be sensed as a warm temperature and the cold as a cold temperature. In this way it is possible to distinguish between the two types of sensation, a thermal comfort sensation and a temperature sensation.

In general, sensory systems in the body become less acute in old age. For example, there are variable losses of vision, hearing, and the sense of smell. At least a part of the decline in sensitivity appears to be due to changes in the nervous system itself, e.g. the degeneration of hair cells in the hearing organs and changes in the auditory nerves lead to losses in hearing acuity. Less is known about age-related changes in warmth and cold sensation and other functions, pain, touch, pressure, and vibration, which are sensed in the skin. It is known, however, that cold receptors in the skin of primates such as the ape are highly dependent for optimum function on a good oxygen- supply. In old age the vascular supply to skin tissues is reduced and the number and sensitivity of functioning nerve cells may alter. Both of these factors could influence the efficiency of thermal perception.

One way to test this is to measure the ability to discriminate between objects heated to different temperatures. The results of tests of this sort applied to young adults and elderly people have shown that whereas nearly all the young people could discriminate differences of about 1 °C between the two objects, the older group usually were not able to match this. Often an elderly person could not discriminate differences in temperature as great as 4 °C. On the face of it, the old perform much worse than young in tests of ability to perceive temperature differences. But there could be an alternative explanation. Perhaps the older people were less confident in reporting different sensations of temperature rather than being less capable of detecting the difference. In order to overcome this problem the results can be analysed by a method which takes account of the way decisions are made — signal detection analysis. These tests showed that there were no differences in the criteria upon which the decisions were made by the young and old groups and this indicates that the differences in temperature perception observed were probably a true indication of an age effect.

The practical consequences of poor ability to discriminate temperature differences can be serious. At one extreme the complete loss of temperature perception in one region of the skin may be the result of spinal cord compression or a central intrinsic tumour of the spinal cord. This may be discovered when for example a patient drops a lighted match on the skin and is unaware of being burnt. Blunting of temperature sensation may simply mean that fine differences in temperature cannot be perceived. But a lowered ability to sense the cold may put some old people at risk if they cannot detect a fall in environmental temperature. One study on elderly people aged between 74 and 86 years showed that most responded to cold discomfort by an appropriate action to produce a warmer environment. Some, however, only experienced the feeling of cold at unusually low temperatures and did not regulate the indoor climate so promptly.

BEHAVIOURAL THERMOREGULATION

This aspect has been investigated in elderly and young people in an attempt to understand how temperature preferences are made. Volunteers were asked to sit alone in a temperature-controlled room for a period of three hours and after the first half an hour, during which time the room was kept at a neutral temperature of 19 °C, control of the room temperature was taken over by the volunteer himself. By switching a switch to 'warmer' or 'cooler' the room temperature could be rapidly altered. Many of the elderly performed this task in the same way as the younger people. Others, however, altered the temperature of the room in a very uncontrolled way, allowing the temperature to swing wildly instead of gradually narrowing the temperature down to the preferred level (Plate 9). The lack of precision in adjusting the room temperature suggests that changes in both temperature perception and behavioural response may increase the vulnerability of some elderly people in cold conditions.

CIRCADIAN RHYTHMS

There is another, more subtle, way in which body-temperature control may be altered in the elderly. It is well known that body temperature is not maintained at a constant level throughout the 24 hours each day. There is a well-marked circadian (meaning about 24 hours) rhythm with body temperature falling to a minimum level during sleep at night and rising to a maximum during the daytime. This body-temperature rhythm appears to be synchronized with the sleep–wake cycle but it has been found that there is an inherent difference in the way these two rhythmical changes are generated by 'pacemakers' in the brain. In some cases when a person is deprived of the 'cues' that time these rhythms, the sleep–wake cycle operates free and with a different periodicity to the body-temperature rhythm. When this happens, and it can be induced for example by constant illumination instead of a normal light–dark cycle, then it has been observed that in a cold environment the body temperature decreases significantly. There is already some evidence that desynchronization of the circadian rhythms occurs more frequently with increasing age and if this is so it would help to increase the risk of hypothermia in elderly people by causing body temperature to fall.

VULNERABILITY TO COLD

The efficiency of thermoregulatory processes appears to be reduced in a proportion of the elderly. Fundamentally, this may be traced back to structural and functional changes in the nervous system and to a reduced blood-supply to organs and tissues during ageing. But there is also a higher incidence of degenerative disease in old people which leads to immobility due to joint disorders, decreased capacity of the heart and lungs to sustain exercise, and the occurrence of mild confusional states, all of which encourage a non-responsive attitude to the possible threat of a cold environment. In this situation there is a greater dependency on physiological mechanisms of temperature regulation, but these may also be deficient. Thermoregulation in the elderly does not fail but the potential is diminished. The effector organs themselves, such as muscles and blood vessels, may be less responsive to nervous control. But changes are also likely to occur in the central nervous system. The control mechanism in the brain may be less 'well-tuned', resulting in re-setting or desynchronization and allowing the body temperature to oscillate uncontrolled between wider limits of internal temperature before physiological adjustments are made.

Apart from these inherent physiological changes there are other specific causes of poor thermoregulation. One is the effect of cerebrovascular disease which may affect the function of the temperature-regulating centres of the brain and even in a moderately warm environment lead to hypothermia (and sometimes hyperthermia). An old person who falls down, perhaps as the result of a stroke, may lie in a cold place for many hours before being found and is particularly liable to hypothermia.

Another disorder found more frequently in the older population is myxoedema which is caused by depressed function of the thyroid gland. Slowness and apathy, characteristic of this disorder, may easily be labelled incorrectly as a normal aspect of ageing and go untreated. In a more advanced stage the face appears swollen and the skin thickened, the hair tends to be sparse, speech slow and monotonous, and the voice deep. The hypothyroidism of myxoedema causes a failure of internal heat production and readily leads to a lowering of body temperature and hypothermia. Small amounts of thyroid hormone gradually restore the patient to normal and body temperature rises once more. The majority of patients brought to hospital with hypothermia are suffering from similar serious disorders which have affected the cardiovascular, endocrine, and nervous systems.

The prospect of hypothermia should not therefore be regarded as an inevitable consequence of old age, despite the observations on failing efficiency of the thermoregulatory system. Most elderly people appear to possess a thermoregulatory system adequate to deal with all but the most severe environmental temperatures. There are not large numbers of elderly people living at home in a chronic state of clinical hypothermia. Secondary hypothermia, on the other hand, has an underlying cause which is in many cases treatable and accidental hypothermia can be largely avoided if old people are given proper care and attention. This is not to say, however, that the incidence of hypothermia in the elderly does not increase during the winter. It increases also in younger groups of people. But when the winter in Great Britain is particularly severe there is only a moderate increase in the number of hypothermia cases resulting in elderly people being admitted to hospital.

9

Indoor climate, comfort, and health

At the time of the Industrial Revolution there was a rapid growth of urbanization; more people moved into the towns and the impact of the built-up environment on health became more clearly recognized. The scourge of epidemics in the larger cities made it essential to develop public health measures to deal with the basic hygiene, the lack of sanitary facilities, and the effects of overcrowding. In many of the present so-called advanced industrial countries these problems have largely receded and attention has now turned toward the quality of the indoor environment. Overcrowding and social problems do still exist, of course, and there is still urgent need for improvements in many aspects of housing and social care. We should be even more aware of the enormous impact of poor housing conditions on physical and mental health, and keep these priorities clearly in view.

Modern technological developments have increased the threat of indoor pollution arising from external sources; again, well-recognized and sometimes controlled on a global scale. Chemical pollutants originating from indoor sources have also been identified in recent years, e.g. formaldehyde from particle-board building materials and foamed insulation, asbestos fibres, oxides of sulphur, carbon, and nitrogen, and house dust. Little is known of the possible long-term effects on health of everyday exposure to low levels of these pollutants. The physical environment of the home, the temperature, humidity, noise, lighting levels all have an impact on health and comfort. Many countries have adopted energy-saving measures in the present era of fuel crises in order to save fuel, by improving the thermal insulation of homes. A secondary effect of this is to reduce natural ventilation in buildings and it has created new health problems by increasing indoor humidity and the build-up of the level of pollutants. Perhaps we are developing an environmental paranoia in the absence of essential facts and a clear perspective. But the indoor environment is built primarily to offer refuge and protection from the extremes of outdoors climate, not to impose a hostile, polluted atmosphere fraught with potential dangers to the health of its occupants.

Temperature in the home is one of the physical characteristics which we care strongly about. Temperature is normally readily perceived and acted upon to achieve thermal comfort. It is the temperature or range of temperatures considered necessary for thermal comfort that is important to most people. Another range of temperatures concerns the levels that may actually pose a threat to health. The higher and lower limits of this range are, however, very difficult to establish with any reasonable degree of accuracy. There are naturally marked differences between different people but three groups stand out as being particularly vulnerable to the effects of indoor climate: the very young, the disabled, and the elderly. It is these groups who are usually obliged to spend most of their time, if not live permanently indoors, and are therefore much more at risk from the effects of low or high indoor temperatures as well as other climatic factors.

TEMPERATURE STANDARDS FOR HOUSING

It has been considered a principal requirement of housing that the design and construction should embody certain environmental health standards. Legislation exists with the aim of maintaining or improving the environment in existing habitations and to limit the building of new housing which is substandard. That standards set down by law sometimes bear little relationship to modern requirements is often due to the cumbersome procedures which are not capable of being quickly updated in accord with technological advances and often also to the lack of reliable information on the effects of different environmental conditions. On the other hand, there is also the basic freedom of choice for people to live in their own dwellings without interference and enforcement of such standards.

Two early studies published by the U.K. Ministry of Works in 1945 and 1953 recommended that for living rooms there should be an equivalent temperature of 17–20 °C, for bedrooms and passageways 10–13 °C, and kitchens 16 °C. The influential Parker Morris Committee report on standards of design and equipment in family dwellings in *Homes for today and tomorrow* (first published by H.M.S.O. in 1961 and since republished many times) came to rather similar conclusions. It recommended that heating installations should be capable of heating the kitchen and areas used for circulation by the occupants to 12.8 °C and the living room to 18.3 °C when outside temperature was –1.1 °C. The Parker Morris report also noted that since 1945 it had by no means been the usual practice for a house to be designed for reasonably low heat loss. Weather stripping of external doors, which can save a large quantity of heat at low cost, was rare, and so were measures to ensure the absence of draughts. In 1970 the Institute of Heating and Ventilating Engineers gave a new set of figures for indoor temperature. Living-room temperatures were recommended to be maintained at 21 °C compared with 18 °C in their 1955 guide and bedrooms at 18 °C instead of 10 °C, and this was basically confirmed by the British Standards Code published in 1977.

Evidence about indoor temperatures in the United States is generally in favour of warmer conditions. The American Society of Heating, Refrigeration and Air-Conditioning in 1972 recommended an indoor temperature of 24 °C but in 1974, 20 °C 'in the interests of fuel saving'.

Hypothermia: the facts

TEMPERATURE REQUIREMENTS OF THE ELDERLY

In many industrialized societies in Europe there is an increasing proportion of elderly people living in the community, up to 20 per cent of the total population in retirement from active employment, including a growing number of those aged over 65 years. For the elderly, control of the quality of the indoor environment is of primary importance. A survey carried out in 1976 by the Office of Population Censuses and Surveys found that in England, 90 per cent of the elderly people over 65 years lived in their own or own family's accommodation, and only 8 per cent in sheltered accommodation or in old people's homes. A high proportion of elderly people therefore control their own environment. Old people in Great Britain tend to live in the oldest dwellings and are often less able to maintain their property and to pay for repairs. The freedom to exert control over their own environment clearly does not match their ability to take action when needed. Old people with less efficient temperature regulation and behavioural response to cold may live in less well-maintained properties which are cold indoors in the winter. A high percentage of the elderly occupy houses in residential areas characterized by unfavourable environmental conditions whereas younger, more energetic people are more able to afford to move out of these surroundings to environmentally favourable areas.

It is recognized that older people prefer a temperature for comfort which is higher than that for younger and more active people. This is entirely consistent with the lower metabolic heat-production of sedentary people or those who are immobilized. A Ministry of Housing circular in 1969 proposed that, in accommodation especially designed for old people, heating equipment should be installed which should be capable of providing a minimum temperature of 21 °C when the outside temperature was –1 °C. This appears to set a reasonable temperature standard for the houses of elderly people. Individual preferences will, of course, not always be met by this guideline and some elderly people prefer a lower temperature; some deliberately choose to live in lower temperatures in order to reduce fuel bills, and others are unaware that the temperature is low. In addition some elderly people, especially those with low thyroid activity, do not feel comfortable even in a room at 21 °C, which they consider too cold. Investigations on the thermal comfort of healthy elderly people show that there is a standard deviation of nearly 3 °C about the mean comfort temperature of 21 °C, which suggests that most elderly people will be satisfied with an indoor climate between 18 °C and 24 °C (given that they are sedentary and wearing 1 clo of clothing).

Indoor temperature

Temperatures in the houses of elderly people in Great Britain, both in the public and private sectors, appear to be generally low in the winter, as several surveys have found. A major investigation of indoor temperatures of the dwellings of those aged 65 years and above was made in the winter of 1972-3.†
This was a random national survey conducted under well-controlled conditions in representative houses in different towns, from Scotland to Cornwall. Indoor temperatures were measured in the morning (8-10 a.m.) and early evening (4-6 p.m.) and it was found that the median temperatures in the morning were between 14 °C and 18 °C and in the evening between 16 °C and 20 °C. In fact, 91 per cent of the living-room temperatures were below the 20 °C standard in the morning and 71 per cent below that temperature in the evening.

Another standard with which to compare these figures is that laid down for temperatures in places of work, the 1963 Offices, Shops and Railway Premises Act. It states that 'where a substantial proportion of work done does not involve severe

† *British Medical Journal* 1, 200 (1973).

physical effort, a room temperature of less than 16 °C after the first hour shall not be deemed a reasonable temperature'. The 1963 Act recommended this lower temperature for workers who, though sedentary, would have a higher metabolic heat production than a relatively immobile elderly person. Yet 54 per cent of the sample of old people in fact had room temperatures in the morning at or below 16 °C and 10 per cent had living-room (a.m.) temperatures at or below 12 °C. The average outdoor temperatures were 7 °C in the morning and 8 °C in the evening so the 1972–3 winter could be considered slightly milder than normal. The national survey does not therefore give a true picture of indoor conditions in severe winter conditions in Great Britain. In a cold week in winter, with outside temperatures dropping below zero, one investigation showed that for the population as a whole, 32 per cent of bedrooms were below 12 °C.

An extensive general survey of domestic dwelling temperatures in U.K. was also carried out by the Building Research Establishment, Department of Environment during February and March 1978, when spot measurements of wet and dry bulb temperatures (see p. 18) were made in each room of 1000 homes nationwide. This included 25 per cent of people who were 65 years of age or more and this elderly sub-sample may therefore be set in a wider, national context. The average temperature of all the dwellings, when average outside temperature was 6–7 °C, was 15.8 °C, with living rooms 18.3 °C, kitchens 16.7 °C, and the warmest bedroom 15.2 °C. Dwellings occupied by elderly people were 0.6 °C cooler than the average.

Of the total sample, about half the homes had no central heating and half had full or partial central heating. Centrally heated houses ran 3 °C warmer on average than non-centrally heated, but the difference in mean temperatures in the living rooms between the two categories was only 1.5 °C dry bulb. Dwellings built since 1970 were found to be 3 °C warmer than those built before 1914. Since the dwellings of the elderly were only 0.6 °C cooler than the average for the whole population it would suggest that a proportion of the elderly in 1978 were living in centrally heated or partially centrally heated accommodation and that not all were living in ancient housing stock. Or alternatively, it might mean that the elderly were paying for extra fuel to maintain the temperature in non-centrally heated, old houses. One striking difference which would more than compensate for the 0.6 °C cooler temperature was the extra clothing worn by elderly people. The mean clothing insulation for men and women under 25 years was 0.74 clo and 0.73 clo respectively. For the elderly it was 0.97 clo for men and 0.90 clo for women. Since + 0.1 clo would be expected to correspond to a change of − 0.6 °C in preferred comfort temperature, and assuming that activity levels and the sensation of comfort did not differ widely between the two groups in this instance, then the greater amount of clothing worn by the elderly would be equivalent to actually maintaining the house temperature + 0.6 °C higher than the younger adults.

Temperature requirements for comfort

'Too cold for comfort' expresses dissatisfaction with the thermal environment in a way which is broadly understood by most people. This judgement is arrived at by a complicated process involving sensing the environment through thermal receptors in the skin, the initiation of physiological responses such as shivering or sweating, and psychological factors which help in forming an impression of pleasantness or otherwise. The sense of thermal comfort or discomfort is a subjective sensation which, under controlled conditions, can be evoked consistently to correspond to a given 'rating' scale, and can be used almost as a human instrument to measure the thermal suitability of an environment. A knowledge of the average thermal comfort sensation is of value to building engineers for improving building design and improving standards of heating and insulation. Apart from the thermal characteristics of the environment itself (temperature, radiant temperature, air movement, humidity) the sense of thermal comfort is affected by a person's activity level (metabolic rate) and the clothing worn (the 'clo' insulation value). It is therefore difficult to generalize about temperature requirements for comfort in the home unless the amount of activity (sitting, standing, walking etc.) and the amount of clothing are specified. Furthermore, to inquire about a person's thermal comfort in his own home is to receive a reply often biased by personal circumstances which may have little to do with the sensation of comfort.

One way of removing this bias is to investigate impressions of thermal comfort in unfamiliar surroundings such as a controlled-temperature climatic chamber. This has been done many times in order to investigate different aspects of human thermal comfort sensations. In tests to compare thermal comfort in controlled temperature environments and comparable activity and clothing levels it has generally been found that for healthy subjects there is little difference between the young and old. In Fig. 2 the predicted thermal comfort sensation for sedentary people is shown for different indoor temperatures and for different amounts of clothing worn. For example, it shows that a sedentary person wearing 1 clo of clothing, equivalent to a business suit for a man or a long thick dress and cardigan for a woman, would feel comfortable in a temperature of 21 °C. A typical list of comfort conditions is given in Table 7.

It should be emphasized that the similarity in comfort requirements between young adults and old people only applies if their activity and amount of clothing worn is the same. It also only applies to people who are healthy and show no gross physiological differences. For example, an elderly person who is hypothyroid may feel cold even in a temperature of 24 °C when the normal elderly person would feel that 21 °C

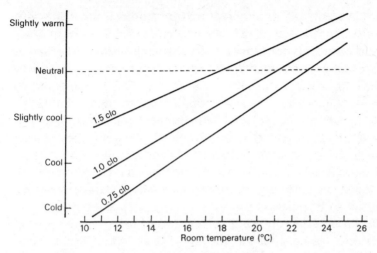

Fig. 2 Thermal comfort of sedentary adults wearing 1 clo of clothing insulation in different dry bulb temperature environments (50 per cent relative humidity and air movement 0.1 metres per second in all environments). Activity is equivalent to category 2 in Table 7 below.

Table 7. *Typical average winter thermal comfort temperature for healthy elderly and young adults in Great Britain*

Activity	Clothing	Room temperature† (°C)
1. Sitting at complete rest	Normal indoor winter wear (1 clo)	23.2
	Heavy indoor winter wear (1.5 clo)	20.6
2. Sitting reading with occasional light activity	Normal indoor winter wear	21.1
	Heavy indoor winter wear	18.4
3. Light domestic work	Normal indoor winter wear	19.8
	Heavy indoor winter wear	16.2

† dry bulb temperature, 50 per cent relative humidity, air movement minimal at 0.1 metres per second

was comfortable. Similarly, there is a small proportion of elderly people who are unable to detect differences in temperature very accurately (see p. 100) and they often prefer an unusually low room temperature.

113

Hypothermia: the facts

The mean indoor temperature for comfort is therefore only a guideline to be applied under known conditions of individual behaviour. It is necessary also to distinguish between temperature requirements in different areas of the house. In bedrooms the temperature is usually lower than in living rooms during the day and often requires 'topping up' at night. Similarly, there may be a mixture of temperature requirements, as for example in residential houses for old people where the requirements of the sedentary elderly residents conflict with that of the more active attendant and domestic staff. A survey of elderly people living at home which was undertaken by the Office of Population Censuses and Surveys and published in 1978 found the following percentages of elderly persons who were not warm enough all the time: 7.6 per cent in bed, 8.5 per cent in the living room, and 12 per cent in the kitchen. Many of those not warm enough blamed inadequate heating facilities, usually as the result of not being able to afford them. It also was clear that a high proportion of old people have to use comparatively costly electric fires and that they had no means of heating their bedrooms, bathrooms, passages, and lavatories. The danger here is that many elderly people have to move repeatedly from a warm to a cold environment during the day and this may increase the impact of the cold-stress.

Temperature requirements for health

If there is an effect of indoor climate on health and, more specifically, of low temperatures in the home during winter, it must affect all age groups. Infants, the disabled, and the elderly are recognized as groups at greater risk. The World Health Organization recommended that dwellings inhabited by the aged should be heated, or capable of being heated, some 2 °C higher than for younger people but, as we have seen, the homes of the elderly are, if anything, colder than average.

The problem is to decide what is meant by 'health'. If 'health' is taken to mean the absence of stress such as that produced by thermal discomfort, then we should start by considering the temperatures beyond the comfort range. Again, the World Health Organization have proposed that an indoor air temperature between 18 °C and 24 °C poses little thermal health risk to sedentary people in appropriate clothing (when there was also air movement less than 0.2 metres per second, relative humidity 50 per cent and mean radiant temperature did not differ by more than 2 °C from air temperature).

Another way of approaching this problem takes into account the fact that people who are ill, will, on the whole, require a warmer environment during the winter, and that limiting low-temperature conditions in the home are the lowest ambient temperatures in which *healthy* people can maintain thermal balance. If a computed model of human thermoregulation is used to calculate these temperatures it predicts that a sedentary, healthy adult wearing 1 clo of insulation could maintain thermal balance and normal deep-body temperature for at least seven hours in an environmental temperature of 5 °C. However, in the absence of shivering and no compensation from increased vasoconstriction (e.g. the extreme case of lack of physiological control in a few elderly people) body temperature would drop to a hypothermic level within seven hours at an ambient temperature of 10 °C and to the same level within four hours at a temperature of 5 °C.

It is probable that indoor temperatures between the limiting condition of 5 °C for those with poor thermoregulatory capacity and 18 °C, the lower limit of thermal comfort for sedentary people, can still impose a threat to health. Presumably, the lower the indoor temperature and the longer the time of exposure, the greater will be the effect. The elderly and young and those disabled or with illness are the groups most at risk. Cold-induced diseases may take many forms. Heart and lung function can be affected by a fall in deep-body temperature even if this is not a profound decrease to the hypothermic level. Cardiovascular reflexes can be initiated by cold air on the face and this can result in a slowing of heart-rate and changes in blood-pressure which might precipitate illness. The effect of cold air on the respiratory passages is to damp down the action of the small hairlike processes called cilia which sweep contaminants away from being absorbed in the lungs. It is frequently observed, though we do not know the precise mechanism, that respiratory infections are often preceded by a chill after getting cold.

Table 8. *Environmental temperatures to provide adequate warmth for babies nursed unclothed in an incubator or clothed and wrapped up in a cot in a draught-free room*

	Birthweight		
	1 kg	2 kg	3 kg
Temperature (°C)	Age of baby		
Incubator			
34–35	1–10 days		
33–34	10 days–3 weeks	1–2 days	
32–33	3 weeks–5 weeks	2 days–3 weeks	1–2 days
Up to 33	After 5 weeks	After 3 weeks	After 2 days
Room			
25–27		1 day–1 week	1 day
21–24		After 1 week	After 1 day

TEMPERATURE REQUIREMENTS OF CHILDREN

The need to provide a thermally neutral environment for newborn babies is particularly critical during the first few weeks of life. A room which is intolerably warm for an adult may be too cold for a young baby. 21–24 °C provides neutral conditions for a full-term cot-nursed baby more than two days old, but for a 1 kg baby in the first few days of life a temperature of 31 °C would be required if the baby was nursed in a cot instead of an incubator. Environmental temperatures to provide warmth for full-term and pre-term babies in cots or incubators are given in Table 8. These are estimates of average temperatures with a range up to 4 °C about the average. For babies in incubators the tolerance is much smaller with a range of temperatures not more than 1–2 °C.

The thermal comfort of children has been studied extensively in the schoolroom, particularly in secondary-school children. There appears to be little difference in temperature preferences of adults and children. Young children have a higher metabolism than adults, have different restrictions or lack of restrictions on clothing, and engage in different and usually more energetic activities. All of these factors will affect a spot-check of thermal comfort. Thermal comfort in children is related to relatively high levels of activity.

10

Treatment and prevention

There is no single, unified method for treating hypothermia nor for that matter a simple set of measures which can be used to ensure that it does not occur. The rule is that each case should be treated individually. Different types of hypothermia call for different methods of treatment, which is perhaps obvious from the description of various categories given in this book. The trauma of sudden immersion in cold water resulting in the rapid development of severe hypothermia requires a different approach than the slow, chronic types of hypothermia which may sometimes be unmasked in disabled or frail elderly people living in cold surroundings. There is still debate as to the most effective methods of treatment in different situations. Should the rewarming procedure be active or passive? That is, should full resuscitation procedures including transfusions of warmth in one form or another be immediately applied, or should there be a gentler, passive warming in a warm room allowing the patient to warm up gradually and spontaneously? The principles underlying the management of hypothermia are much better understood than they were 20 years ago when 'urban' hypothermia was first recognized, and treatment is now more successful, particularly in babies.

Studies on the prognosis of patients admitted to hospital with hypothermia often show that it is some underlying disease which is more important in determining survival than, for example, the method of rewarming used. Recovery also usually depends on the severity and duration of exposure to cold as well as on general health before the incident. The prospect of normal recovery, barring other complications, is particularly good in young adults if the deep-body temperature does not fall lower than 32 °C. If body temperature falls as low as 26 °C intensive care may enable many young victims to recover, but there is the added possibility that some form of lasting tissue damage may occur particularly of the nervous system. When body temperature falls below 26 °C, a large proportion of casualties, especially old people, do not recover from severe hypothermia in spite of treatment.

TREATMENT OF HYPOTHERMIA OUTDOORS

There are three general principles which should be borne in mind in managing a case of accidental hypothermia discovered outdoors. First, the victim will stay chilled or will cool further unless some form of warmth or protection is applied. Secondly, whatever the severity of the condition the casualty should be seen as soon as possible by a doctor. Lastly, if the body temperature is less than 32 °C the patient must be considered a medical emergency and taken immediately to hospital, not only in order to treat the hypothermia but to remove him or her from the stressful environment.

Cardiopulmonary resuscitation

Essential first-aid life-saving procedures are familiar to most people:
  (1) clearing the upper respiratory passages;
  (2) mouth-to-mouth ventilation of the lungs;
  (3) external compression of the chest to stimulate the heart's action.
These are required in hypothermic victims who reach the point at which the heart stops and breathing ceases, but heart massage should be started at half the normal rate. The difficult predicament which faces the would-be rescuer, however, is to decide whether the patient still has a heartbeat

119

and respiration, perhaps almost imperceptible, or whether these functions have ceased and it is appropriate to start cardiopulmonary resuscitation with external chest compression, which itself carries the special risk of inducing ventricular fibrillation in a victim of hypothermia. This decision is best left to a trained medical attendant if one is available. Once resuscitation procedures are started they must be continued until the victim has recovered or until artificial support to vital functions can be given. If there is any respiration at all it is safe to assume that the heart is beating and therefore it is not advisable to start cardiopulmonary resuscitation.

Rewarming

Rewarming procedures in outdoor locations have been discussed in Chapter 6; facilities for rewarming are more often than not extremely limited and the art of compromise frequently called upon. A low-reading thermometer is unlikely to be available. If the body temperature can be taken then the rectal temperature is the most useful; it is not usually possible to take urine temperature, especially if the patient is dehydrated. In the field situation it is often impossible to do more than take simple measures to protect the victim from the environment in an attempt to prevent body temperature falling any further. Rewarming procedures usually accomplish the final aim of raising body temperature but this is a slow process unless special equipment is available. The most efficient method of rapid rewarming is to immerse the patient in a hot bath at 41–45 °C, keeping his limbs out of the water. Cooling of vital organs such as the heart and brain occurs much more readily during the 'after-drop' phase if the limb blood vessels are rapidly dilated during rewarming. The risk of rewarming-shock is reduced when only the trunk is rewarmed first.

The after-drop in body temperature during rewarming

occurs in most cases of hypothermia when the victim has been recovered from water. Collapse may occur soon after the patient is rescued. This is not solely due to the effects of the after-drop but also to removal of the 'hydrostatic squeeze', i.e. of external water pressure acting on the superficial blood vessels. When a person is first removed from the water, therefore, blood pressure may drop and cause him to collapse.

One word of caution should be added to these guidelines concerning rewarming. If there is good reason to believe that the signs and symptoms observed are due to hypothermia then rewarming procedures may be undertaken. But they should not be proceeded with if there is doubt that the victim is hypothermic or if the semi-conscious patient may be suffering from some underlying disorder.

TREATMENT OF HYPOTHERMIA IN HOSPITAL

Infants

When there is a risk of hypothermia the baby's temperature should be carefully maintained and changes of temperature followed. In hospital the temperature-controlled incubator controls the baby's environment. But a single measurement such as air temperature in an incubator describes only one aspect of his thermal environment. Heat is lost also by evaporation and radiation and this will not be directly measured simply by dry bulb temperature of the air.

The safest method of treatment is to place the baby in an environment which is a few degrees warmer than deep-body temperature and, by blocking the main avenues of heat loss, to allow the baby to warm up by means of his metabolic heat production. By adjusting the incubator temperature upwards at intervals the body temperature may be allowed to rise by about 1 °C every hour. If hypothermia has been present for more than about six hours it is the usual practice to raise body temperature rather more slowly than this by gradual manipulation of the incubator temperature.

Adults

Most adults will warm spontaneously and once shivering has started warming will be quite fast, at the rate of 1–2 °C per hour. Three methods of rewarming have been used.

1. *Active external warming.* The surface of the body can be warmed by hot-water bottles (well-wrapped to ensure the skin is not burned) or an electric blanket, by immersion in a bath of warm water, or by lying on a mattress filled with warm water.

2. *Active internal warming.* This can be accomplished by various clinical methods, usually in an intensive care unit where equipment for maintaining the vital functions of the patient are readily available. Blood can be warmed outside the body when it is passed through the heat exchanger of a heart-lung machine and then recirculated back into the patient. Irrigation of warm fluids into the abdominal cavity is another method of providing extra heat to the trunk of the body. Warm-air mixtures to breathe are also sometimes used, though there is no clear evidence that this helps to raise body temperature substantially.

3. *Passive warming.* This method embodies a more conservative approach designed to disturb the patient as little as possible and to allow slow rewarming. Careful insulation by extra clothes or bedclothes in a warm room between 24 °C and 29 °C is usually effective in reducing heat loss and allowing body temperature to rise through the patient's own internal heat production.

Some centres claim considerable success in resuscitating hypothermic patients in an intensive-care unit by a combination of active measures. These include rapid warming of the trunk by a heat cradle to raise body temperature at the rate of 0.5–1 °C per hour, warmed intravenous fluids, and intermittent positive-pressure ventilation with the administration of oxygen.

The elderly

It is still the general practice to rewarm elderly hypothermic patients gradually by passive procedures. There is less agreement about using this method in old people who are severely hypothermic and who require some nursing in intensive-care conditions. Mildly hypothermic elderly patients are usually 'barrier' nursed in a cubicle and placed on a ripple mattress and covered with good insulation in a warm room. The aim is to raise body temperature by not more than 0.5 °C per hour and on this regime most mildly hypothermic patients return to normal temperatures within 12 hours.

SOCIAL CARE AND THE PREVENTION OF HYPOTHERMIA IN THE ELDERLY

The vulnerability of the elderly population to cold is now well-recognized and an excellent booklet *Keeping warm in winter*, produced by the D.H.S.S. in 1972 (published by H.M.S.O), is one of many social services guides which exist to help elderly people improve their thermal living conditions in the U.K. Similarly, practical advice on protective measures against hypothermia is available in a pamphlet published by the U.S. Department of Health, Education and Welfare in 1978 entitled 'Accidental hypothermia: a winter hazard for the old'. Public attention inevitably focuses on hypothermia in the old during the winter, but it is important not to lose sight of the much larger general problem of the elderly living in uncomfortably cold surroundings. There is often an association between low deep-body temperature in old people and adverse social circumstances; see Plate 10. Some of the earlier social surveys in Great Britain indicate that the population particularly at risk from cold are those over 75 years of age, isolated and living alone and often those eligible for supplementary benefits. The temperature of a room, and humidity, in which an old person lives is of considerable importance and unhappily in times of inflation and escalating costs of fuel old people tend to cut down on heating.

The principles of preventing the development of hypothermia should be recognized by those attending the elderly who are living in isolated social circumstances. Of primary importance is the early detection of incipient hypothermia by maintaining regular surveillance. Special attention should be given to the elderly living in cold accommodation and with low body temperatures, measured by urine temperature, even though they may not complain of cold. From the point of view of those in direct contact with the elderly in the community, certain key factors serve to alert the visitor. They are (1) low temperatures in the living room and bedroom and lack of potential heating sources; (2) poverty combined with social isolation; and (3) a general deterioration in well-being and body functions. Food is an important consideration, for a small energy intake can affect the amount of heat the body produces internally. If extra warmth is required during a cold winter spell one of the inevitable choices is to spend less on food. The delivery of hot meals (meals-on-wheels service) to the elderly is a vital means of helping to prevent hypothermia.

In an attempt to conserve resources by reducing expenditure on increasingly expensive fuel some old people may keep the temperature of their living spaces too low for comfort, which may encourage a state of 'voluntary' hypothermia. Provision of adequate space heating in the house is perhaps the most important preventative measure. Public provision of support for the elderly in need of extra heating usually takes the form of three measures: (1) payment for fuel and schemes developed by local-authority social-services departments for phasing payments; (2) payment for heating appliances such as electric fires and electric blankets; and (3) the provision of house insulation. Unfortunately most elderly people do not take advantage of existing benefits. In 1972 it was found that three out of four elderly people were unaware of the extent of supplementary benefits, and only a small proportion (11 per cent) were actually receiving extra heating allowances. Recent estimates indicate that the situation has changed in the last decade and that more than 60 per cent of those entitled are now receiving extra heating allowances. It should be remembered, however, that perhaps more than 70 per cent of those over pensionable age are not entitled to supplementary benefits or help with heating though many may be only marginally more affluent than those who are entitled.

Measures designed to protect the elderly from cold environments by providing extra clothing present problems. Although adequate clothing is a necessary defence against cold and provides a simple and relatively cheap method for controlling an individual's micro-climate, the physics of clothing argues against it as a means of efficiently preventing hypothermia. In cold conditions, clothing becomes efficient in controlling heat loss when metabolic heat production is high. In the hypothermic patient metabolic heat production is low and the thickness of clothing has to be increased out of all proportion in order to provide sufficient insulation to prevent heat loss. Many instances of hypothermia, have, moreover, been reported in elderly persons who were in bed apparently well covered with clothes and good external insulation. The use of clothing also depends on the behavioural response to the sensation of cold and the mobility to put clothing on without difficulty. Both of these functions are often impaired to some extent in the elderly. Garments manufactured on the 'space suit' principle with metallized surfaces are commercially available but are usually expensive. The technology for providing heated clothing is also available but is also expensive. These garments carry added disadvantages. Some form of safe electrical heating is usually guaranteed but it is unsafe to have electric leads attached to garments worn by elderly people for they may trip over the leads. Furthermore, heated garments have the added disadvantages of encouraging a static state and inhibiting activity in the old.

The technical implications of improving the thermal efficiency of dwellings occupied by the elderly have been explored as a method for implementing first-aid constructional work at relatively low cost. Such work on houses with the installation of new, alternative, or additional heating appliances, draught exclusion, and better insulation is reported by elderly people to have led to an improvement in their thermal comfort. But piecemeal improvements are not ideal and may not even achieve the minimum temperature levels recommended by the Parker Morris report on standards.

The feasibility of installing warning systems to monitor body and room temperature in order to detect dangerously low temperatures has often been considered. The remote monitoring of room temperature in selected cases would be possible but the problem of defining minimum temperature standards for all old people remains. The difficulty inherent in a system which the occupant can use to monitor temperature is that his capacity to act on adverse information may not be adequate. Some elderly people experience difficulty in monitoring and controlling heating systems even in their own home.

11

Future trends

Extreme cold, as we have seen, can be considered from two opposite points of view. It can be harnessed for our own benefit in many ways: to prevent pain, to give an added margin of safety in open-heart operations, to preserve human cells, and perhaps one day to prolong useful life. Cryobiology, the science of freezing living cells and tissues, has added a further dimension to the uses to which cold can be put for the help of mankind. It has already brought incalculable benefit by allowing prolonged preservation of red blood cells and the storage of blood in blood banks, and the storage of animal sperm for cattle breeding. It also provides the potential for preserving organs to be used in transplant surgery, though here we are approaching the frontiers of knowledge where further progress is still shrouded in controversy. But cold is also a killer which threatens the lives of individuals overwhelmed by the outdoor elements and has insidious effects on the neonate, the disabled, and the aged even in protected environments.

There now exists a substantial body of information on the clinical condition of hypothermia, where it is found, its particular dangers and its treatment. Neonatal hypothermia is one aspect which has received special priority and the conditions under which it occurs and its management are well understood. The outlook of severe hypothermia in the newborn is extremely poor, especially in light-for-dates and preterm infants, so it is essential to adopt sound preventative measures to avoid chilling at and for some weeks after birth. Warm ambient temperatures in the nursery and, if necessary, the use of an incubator is the key to successful prevention. Vigilance by responsible adults is also required. Hypothermia is known to occur even in the tropics in infants who are severely malnourished and have defective internal heat production and poor body insulation.

The outstanding problems associated with accidental hypothermia in adults are well within the bounds of possible solution. Diagnosis has undoubtedly improved with the increasing availability of the low-reading clinical thermometer, but existing non-invasive methods of measuring deep-body temperature are often slow and cumbersome. Advances in thermometry are continuously being made and it is possible to foresee the development of devices which will accurately measure deep-body temperature from placement on the skin surface. Prototypes of such devices are already in existence. The main clinical problems generally concern intensive-care techniques, where improvements will enable more success in resuscitating severe cases of hypothermia.

There will always be a proportion of cases of secondary hypothermia where the underlying medical condition is severe enough to defy all efforts at resuscitation. Inevitably there will also be new combinations of drugs taken in overdose amounts, which are found to have the property of lowering body temperature to hypothermic levels.

The treatment of accidental hypothermia arising from immersion exposure or therapeutic accidents is not simply a matter of applying whatever method of rewarming is readily available. More research is needed into the use of surface and core warming, particularly when it is applied to the elderly patient who may be less capable of benefiting from aggressive intensive-care techniques. The dilemma of whether or not to commence cardiopulmonary resuscitation in the severely hypothermic victim of exposure is one which requires a degree of clinical acumen at the scene and an appraisal of local circumstances. The danger of precipitating ventricular fibrillation in an hypothermic patient by rough handling is very real. There are many instances, however, where hypothermic patients have appeared to possess no vital signs, were not actively resuscitated, and yet eventually revived.

The 'old and cold' is a theme which frequently attracts public concern in Great Britain and there is no denying the long-term need to improve the thermal standards of the housing stock in this country. We have a much clearer idea now of how the condition of hypothermia in the elderly can arise. In a proportion of old people the evidence of physiological studies suggests that the ageing thermoregulatory system may be at fault. But it is unusual for this to be the direct and only cause of hypothermia in cold conditions, though it will of course increase vulnerability.

Secondary hypothermia, resulting from some underlying medical condition, appears to be the most common reason for elderly people to be admitted to hospital with hypothermia. This form of hypothermia may sometimes be unconnected with low temperatures in the dwelling. Statistics from hospital admissions have not yet been analysed to show the extent to which deep-body temperature falls as the result of a heart attack or stroke. The difficulty is to obtain reliable information on this aspect in an emergency. Another important effect of cold is that it induces an additional 'strain' on the heart and circulation and this may take a significant toll of unfit elderly people during the winter. Even in younger adults breathing cold air produces benign changes in the action of the heart, changes which may be harmful in the elderly or unfit. One post-coronary rehabilitation jogging programme could not be continued in winter without some protection against breathing cold air. Exercise combined with cold-exposure of the face and extremities produces a number of cardiovascular and blood-pressure responses which increase the strain on the heart and circulation. A secondary effect of a failing circulation in these conditions is likely to be hypothermia.

Finally, there is also a significant number of cases of accidental hypothermia in the elderly caused by a fall at night in a cold bedroom or a slip on the ice in the garden, which often leads to immobilization and excessive cold-exposure. In recent years there has been greater medical and sociological awareness of the problem, and this combined with the widely-practised 'good-neighbour' policy has helped to limit the number of cases of elderly persons suffering from hypothermia. Eventually, warm dwellings achieved within the financial capability of elderly people must be the goal of all those aiming to reduce the incidence of urban hypothermia to an absolute minimum.

Cold does not in itself cure disease nor increase healing but it is capable of reducing metabolic processes and local blood-flow and can to some extent reduce certain stresses on the body. Inflammatory processes can be slowed and pain alleviated so that the intensity of their combined effects may be controlled. Induced hypothermia has become an established part of anaesthetic technique in many present-day surgical operations. In cardiac surgery, it is becoming even more widely used in the large number of operations likely to be undertaken by the relatively new technique of vein grafting in coronary vascular surgery. One of the disadvantages of using a heart–lung machine (cardiopulmonary bypass) is that blood-pressure tends to be kept at a steady low level, without the normal fluctuating systolic and diastolic pressure, and this can cause damage to vital organs such as the brain. Hypothermia is used to minimize possible damage and to allow a greater margin of safety. Several studies have reported better protection from organ damage using induced hypothermia, and a significant improvement in function when the heart is restarted. Nowadays the technique may involve keeping the heart at a temperature below 20 °C by perfusion with cold blood, or local cooling applied directly to the heart, or both, and it is routinely used for babies undergoing heart surgery.

The ability to freeze, store, and thaw red blood cells enables large-scale blood transfusions to be based on blood banks. The development of cold technology has extended the shelf-life of fresh blood, which could previously be stored in useable form for only three weeks. It may be kept now for an almost indefinite period. Thousands of units of blood were once wasted each year because of spoilage, but now whole blood and almost all of its component cells can be stored frozen for long periods. Centres such as the European Blood Bank have been established for freezing and storing new or rare types of blood components.

Patients who are transfused with refrigerated blood readily become hypothermic, especially if cold transfusions are given in air-conditioned operating rooms during operations which expose deep-body cavities. Core temperatures can quite easily fall to 30 °C or below and this is a serious complication because blood coagulation is impaired, the affinity of the blood for oxygen is changed, metabolism is slowed and there may be cardiotoxic effects. Much research has gone into investigating the methods of warming blood obtained from cold storage by placing it in tubes which are coiled in warm water or by warming with electric heating-plates and by microwaves. The latter method is the least successful but more studies are still needed to find the best method for rapidly warming blood from its storage temperature of 4 °C to 37 °C for transfusion.

The other notable achievement in cryobiology is in the field of reproductive medicine. Most of Britain's dairy cattle are now produced by artificial insemination with sperm which has been stored frozen. There is little difficulty in preserving human sperm for long periods at low temperatures. Recent advances have enabled scientists to freeze and store both animal and human fertilized eggs. Mouse embryos have been stored for as long as eight months in liquid nitrogen after which time they have been thawed, implanted in a foster mother, and eventually a normal delivery of live young achieved in a high proportion of animals. The future will surely see a gradual and ever-widening application of this technique of storing fertilized eggs with subsequent implantation. In humans, the use of these techniques has brought special difficulties. The problem is that technology seems to have outstripped existing ethics.

Hypothermia: the facts

Fascination with the preservative power of freezing ante-dates even the discovery in 1799 of the frozen bodies of mammoths in Siberia. Now almost all types of individual cells can, by using different cooling and warming techniques, be frozen, stored, and actually brought back to life. The process can also be applied successfully to small amounts of homogeneous tissue such as the corneas of eyes, cartilage, and smooth muscle. It is easy to visualize the possibility of freeze-storing bone marrow transplants for the treatment of anaemias and leukaemias, and pancreatic tissue transplants for diabetic patients.

There is as yet little experimental evidence that it is possible to freeze whole organs to a stabilizing temperature with eventual recovery for transplantation. Efforts were directed toward this end in the 1970s with the aim of setting up organ banks for future patients suffering from failure of the heart, kidneys, lungs, or liver. But in subsequent years it was shown that the preservation of transplant organs by freezing was even more difficult than the problem of transplant rejection. Cryobiologists have not yet succeeded in storing organs for more than 72 hours. Only with the kidney has there been much success at transplantation on a large scale and even then the storage time is no more than 10 hours. Technical advances can, however, be realistically expected which will enable organs to be stored for longer periods by using perfusates and cryoprotectants.

Freezing life itself has for some time appeared to be an application of cryomedicine within the reach of possibility. But this is really not so. There is no avenue of research at present which offers the hope that this can be accomplished. Individual cells can be preserved but not all cells respond equally well to freezing. No single freezing procedure is adequate to preserve for eventual recovery a complex organ system composed of many different types of cells. The final pragmatic word might be accorded to Professor N. Kurti, an eminent Oxford cryophysicist, commenting on the beliefs of

Future trends

the cryonic societies: 'One may laugh this off, but the sad thing about it is that relatives probably honestly believe that their beloved will be resuscitated, and to give hope where no hope exists is very cruel indeed.'

Index

Index

endocrine system 15, 32
Eskimos 3, 14, 27
evaporation 19, 55

face cooling 116
Fahrenheit scale 27
fatigue 67, 73
fetus, temperature of 53
fever 4, 54, 81
food, specific dynamic action of 19, 93
Four Inns walk 66
freezing 11, 132

gasp reflex 38, 76
glaciation periods 2
greasing 77

heart 37
 attack 32, 129
heat
 balance 17
 definition of 16
 energy 16
 exchange 17
 storage 17
heating allowances 125
hibernation 12, 23, 81
hikers 62
homeothermy 12
house insulation 50
housing 105, 110, 126
hypnotics 86
hypoglycaemia 85
hypoglycaemic agents 87
hypothalamus 15, 31, 86
hypothermia
 accidental 33, 60, 104, 128
 definition of 1
 drug-induced 32, 81
 immersion 7, 60, 74
 in elderly 45, 91
 in hospital studies 42
 in infants 58, 121
 in newborn 53
 in surgery 8
 in World War II 8
 management of 69
 mortality from 46
 nature of 30
 prevention of 123
 primary 31

recognition of 59, 64
secondary 31, 104, 129
statistics of 41, 47, 58
therapeutic 4, 33, 81, 130, 189
treatment of 118
urban 41, 45
voluntary 124
hypothyroidism 32, 86, 93, 103, 109, 112

ice age 2
igloos 27
incubators 53, 116, 121
indoor
 pollution 105
 temperatures 106, 108, 114
Institute of Heating and Ventilating Engineers 107
insulation
 clothing 24
 fat 23, 55
insulin 90

joule 16
J-wave 37

kidney function 39

Lakonia 7, 87
light reflex 40
lithium 88
low birthweight 55
lytic drugs 4, 81

mental impairment 36, 39
metabolic
 rate 18, 54, 91, 93
 acidosis 39
metabolism 17, 36
military campaigns 5
mortality
 seasonal 49
 statistics 46, 50
mountain
 disrobing syndrome 73
 sickness 74
mountaineers 72
muscle rigidity 40
myxoedema 103

National Center for Health Statistics (USA) 48

135